The Management of Technological Innovation

The Management of Technological Innovation

An International and Strategic Approach

MARK DODGSON

UNIVERSITY PRESS

OXFORD
UNIVERSITY PRESS

Great Clarendon Street, Oxford OX2 6DP

Oxford University Press is a department of the University of Oxford.
It furthers the University's objective of excellence in research, scholarship,
and education by publishing worldwide in

Oxford New York

Athens Auckland Bangkok Bogotá Buenos Aires Calcutta
Cape Town Chennai Dar es Salaam Delhi Florence Hong Kong Istanbul
Karachi Kuala Lumpur Madrid Melbourne Mexico City Mumbai
Nairobi Paris São Paulo Singapore Taipei Tokyo Toronto Warsaw

with associated companies in Berlin Ibadan

Oxford is a registered trade mark of Oxford University Press
in the UK and in certain other countries

Published in the United States
by Oxford University Press Inc., New York

First published 2000

British Library Cataloguing in Publication Data

Data available

Library of Congress Cataloging in Publication Data

Dodgson, Mark, 1957–
The management of technological innovation: an international and strategic approach/
Mark Dodgson.
p. cm.
Includes bibliographical references.
1. Technological innovations—Management. 2. Research, Industrial—Management.
I. Title.
T173.8. D62 2000 658.4′062—dc21 00–022543

ISBN 0–19–877536–9 (Hbk.)
ISBN 0–19–877535–0 (Pbk.)

1 3 5 7 9 10 8 6 4 2

For Kate

PREFACE

The management of technological innovation (MTI) is one of the most important and challenging aspects of contemporary business. With technological innovation becoming the fundamental driver of competitiveness for firms in a wide variety of business sectors, it is essential that both the tools and techniques and the strategies of MTI are well understood.

This book describes and analyses these tools, techniques, and strategies. It examines them in respect of research and development, new product development, operations and production, technological collaboration, technology strategy, and the commercialization process. It also analyses the high levels of complexity and risk associated with MTI. While MTI is one of the most important aspects of management, it is also one of the most difficult.

Throughout the book, the particular aspects of MTI are discussed in the context of the general environment in which businesses are operating—an environment that is extraordinarily dynamic. The organization and structure of industry and business are being transformed; firms and their technology are increasingly becoming integrated into various networks and systems; management philosophies and practices are altering with increased focus on issues such as learning, knowledge, and trust; the innovation process itself is changing with the use of new electronic media; and globalization is having a greater impact on management practices. All these changes impact upon MTI. Throughout the book a number of case studies are used to illustrate key themes, and a series of 'brief notes' are used to analyse particular issues.

This book aims to differentiate itself from existing books on MTI in a number of ways. First, its breadth of coverage. It examines a broader range of aspects of MTI than the majority of books in the field, moving from the management of basic science to factory production, from consideration of technology strategy to operations management. It considers some issues from a theoretical perspective, others from the practical point of view of 'what is the best way of managing this?'

Secondly, it is international in focus. The vast majority of books on this subject take a US perspective (and a US large-firm perspective at that). However, technology development and diffusion are essentially international activities, and technology-based firms usually have an international focus either through selling products and services overseas, or by working with foreign suppliers and partners. This book uses material revealing best practice from wherever it is found, be it the United States, Europe, or Asia. It is not so long ago that large parts of US industry believed it had lost the competitive battle against Japan (Dertouzos *et al.* 1989), and the World Bank talked of the 'Asian Miracle'. In the late 1990s the US economy flourished, and most of the East Asian economies faced considerable difficulties. History shows us that it is unwise to think of these circumstances as anything but temporary. Permanent, sustained corporate advantage depends upon learning about the best management practices available internationally—found in firms in the UK, the United States, Taiwan, Italy, Korea, Japan, Israel, Australia, or elsewhere—and understanding that there are always better ways of doing things and improving. No one country has a monopoly on good ideas, and it is mistaken to believe that any particular international model of management has all the answers for MTI. This is important, as many of the most pressing challenges for the future of MTI relate to its international management.

Thirdly, as well as examining the activities of large firms, the book discusses MTI in small and medium-sized firms. These firms are important sources and users of technological innovation, and the challenges and opportunities they face are often different from those confronting larger firms.

Fourthly, many of the issues discussed are applicable to both manufacturing and service companies. As we shall see, the boundary between the two is becoming increasingly blurred, and the MTI best practices found in one sector are often successfully transferred to others.

Fifthly, in an area where there seems to be an obsession with using only the most up-to-date material, the book integrates analyses and research findings from the 1950s to the late 1990s. A great deal of insightful and useful research into MTI tends to be ignored because it is not deemed appropriate to present circumstances. An alternative view is held here; much of this older research is relevant, and the book draws on research from the 1950s, 1960s, 1970s, 1980s, and 1990s.

Sixthly, the book is not driven by a desire to provide simple answers to the problems of MTI or to promote a particular management technique. There are no magic recipes for the management of MTI. Managing technology and innovation is a difficult and idiosyncratic process within individual firms. Some of the techniques that have been sold in the past with evangelical zeal, such as business process re-engineering and total quality management, have a role to play in MTI, but they are not the solution. The real picture, as we shall see, is much more complex.

Seventhly, the book takes a strategic approach to MTI. It shows how many of the tools and techniques used in the management of R & D, new product development, operations and production, and technological collaboration are not fully effective unless they are guided by a strategy. Such a strategy has to be informed by an understanding of the broad changes taking place in industry and business, and the book locates MTI in the context of the resulting challenges.

The book is structured into three major sections. In the first, Chapters 1 and 2 define MTI and examine the broad context in which it operates. Chapter 1 examines the specific and more general areas that MTI encompasses. It uses a number of compiled case studies to illustrate the challenges facing technology-based firms. Chapter 2 examines the new challenges of MTI by analysing some of the major contextual and environmental changes occurring in contemporary business and management.

The next part of the book, Chapters 3, 4, 5, and 6, looks at specific issues of MTI, including: R & D, new product development, operations, and technology strategy. A number of approaches, tools, and techniques for managing these areas are described. Chapters 3, 4 and 5 conclude with the view that, while these methods can be highly efficient they will not be effective in building company competitiveness unless they are managed strategically. They provide the basis for the discussion of technology strategy in Chapter 6. This chapter highlights the difficulties in analysing and developing technology strategies, but re-emphasizes its importance.

The final part of the book discusses some important issues for MTI. Chapter 7 addresses technological collaboration—the variety of ways firms cooperatively produce and sell new technologies. Chapter 8 considers the issues of commercializing technological investments, including some aspects of technology transfer. Chapter 9 concludes the book with an examination of the challenges of MTI for the future.

The need for an international perspective to MTI has been brought home to me by my research, but perhaps even more forcibly by my students. My teaching over the past fifteen years or more—in postgraduate degrees in MTI, in undergraduate courses for engineers, and in international MBAs—has brought me into contact with excellent students from all around the world. Their world views, experiences, and future career aspirations are totally international. Their enthusiasm, insights, and internationalism inspired this book. I hope it will reflect their ideals and guide their successors.

ACKNOWLEDGEMENTS

The antecedents of this book lie with that remarkable collection of eclectic academics at the Science and Technology Policy Research Unit (SPRU) at the University of Sussex. Chris Freeman, Keith Pavitt, and Roy Rothwell have inspired many over the years, and I count myself fortunate to include myself among them. I have drawn heavily on the research and teaching at SPRU, and thank particularly Margaret Sharp, David Gann, Paul Gardiner, and Paul Quintas for their contributions to my research and thinking in this area.

Various friends and colleagues have commented on the book at different stages of its gestation. My thanks are extended to Jon Sigurdson, John Mathews, Ed Russell, Mick Cardew-Hall, Mark Matthews, André Morkel, Krystyna Palonka, Oscar Hauptman, and Al Goldberg, all of whom have made very helpful suggestions. Jane Marceau helped significantly with the sections on systems of innovation. John Bessant's comments were as insightful as ever, and reminded me why I had the good sense to write my last book with him. My thanks also to my colleagues at the Australia Asia Management Centre at the Australian National University, particularly Bruce Stening, who took on extra burdens to enable me to find the time to research and to write. My thanks are also extended to Asia Pacific Press and Hilary Walford for editorial assistance, and the ever-efficient Vicki Veness for her help with the manuscript.

CONTENTS

LIST OF FIGURES

LIST OF BOXES

LIST OF TABLES

ABBREVIATIONS

AGV	automated guided vehicle
AIDS	auto immune deficiency syndrome
APEC	Asia Pacific Economic Cooperation Forum
ASIC	application-specific integrated circuit
ATM	automatic teller machine
BTG	British Technology Group
CAD	computer-aided design
CAE	computer-aided engineering
CEO	chief executive officer
CERN	Centre Européenne de Recherche Nucléaire
CHI	Computer Horizons Inc.
CIM	computer integrated manufacturing
CMM	capability maturity module
CNC	computer numerical control
CoPS	complex product and systems
CPU	central processing unit
CTO	chief technology officer
CSIRO	Commonwealth Science and Industry Research Organization
DCF	discounted cash flow
df/i	direct fuel injection
DRAM	dynamic random access memory
EDI	electronic data interchange
EDP	electronic data processing
EIRMA	European Industrial Research Managers Association
EPIC	electronic pre-assembly in the computer
EPLD	electrically programmable logic device
EPOS	electronic point-of-sale
ERP	enterprise resource planning
ETSI	European Telecom Standards Institute
EU	European Union
FMS	flexible manufacturing systems

GATT	General Agreement on Tariffs and Trade
GDP	gross domestic product
GE	General Electric
GM	General Motors
GSM	global system for mobile telecommunications
HP	Hewlett-Packard
HRM	human resource management
ICT	information and communications technology
IPR	intellectual property rights
IRAP	Industrial Research Assistance Program
ISDN	Integrated Services Digital Networks
ISO	International Standards Organization
IT	information technology
JIT	just-in-time
LANS	local area networks
MFN	most-favoured nation
MRP	materials requirement planning
MRPII	manufacturing resource planning
MTI	management of technological innovation
NAFTA	North American Free Trade Agreement
NASA	National Aeronautics and Space Administration
NEC	Nippon Electric Company
NIS	national innovation system
NMCTTC	NASA's Mid-Continent Technology Transfer Center
NPD	new product development
NSF	National Science Foundation
NTBF	new technology-based firm
OEC	Orbital Engineering Company
OECD	Organization for Economic Cooperation Development
OEM	original equipment manufacture
PC	personal computer
PDM	product data management
QA	quality assurance
QC	quality control
R & D	research and development
ROI	return on investment
SEC	Samsung Electronics Corporation
SIN	systems integration and networking
SME	small and medium-sized enterprise
SMF	small and medium-sized firm
SPRU	Science and Technology Policy Research Unit
SVP	senior vice-president
TI	Texas Instruments
TMRBs	Technology Management Review Boards
TQM	total quality management

TRIPs	Trade-Related Aspects of Intellectual Property Rights
UN	United Nations
UNDP	United Nations Development Program
WIPO	World Intellectual Property Organization
WTO	World Trade Organization

CHAPTER

1

What is the Management of Technological Innovation and Why is it Important?

The challenge of management in business firms is to develop and sustain competitive advantages. These competitive advantages allow firms to meet their objectives, be they profit generation, growth, increased market share, or increased employee remuneration and job security. This book examines the ways in which the management of technological innovation contributes to the development and sustainability of competitive advantages.

Firms compete successfully when they offer the new, better, and/or cheaper products and services that their markets and customers require, and that their competitors cannot provide. Competitive advantage therefore derives from the ability to make and do things more cheaply and better, or to make and do new things. It has a relative dimension: competitive advantage derives from the activities of firms compared to those of their competitors. It also has an absolute dimension: there has to be a market for what the firm does. Technological innovation plays a central role both in improving productivity and developing new products and services, and in providing comparative and absolute advantages. Indeed, it is argued here that, as we move towards what is called the 'knowledge economy', technological innovation will become the primary strategy for competition in the twenty-first century. The *Economist* has gone so far as to call innovation the 'industrial religion' (20 Feb. 1999). The management of technological innovation (MTI) is, therefore, a critically important activity.

Technology includes not only tangible artefacts, but also the knowledge that enables it to be developed and used in ways that are useful. In this sense, technology has to deliver replicable functionality. The definition used here, therefore, is that: technology is manifested in new products, processes, and systems, but it includes the knowledge and capabilities needed to deliver functionality that is reproducible.

Innovation is much more than invention, and it includes all the activities encouraging the commercialization of new technologies (Freeman and Soete

1997). The definition used here, therefore, is: innovation includes the scientific, technological, organizational, financial, and business activities leading to the commercial introduction of a new (or improved) product or new (or improved) production process or equipment. These definitions are broad, and encompass a wide range of issues. What, then, is the management of technological innovation? It includes both specific and general issues. The specific issues are depicted in Fig. 1.1.

SPECIFIC AREAS TO BE MANAGED

Centrally important is the management of *research and development*. This includes issues ranging from techniques of technology forecasting and market assessment to organizational questions, such as the centralization or decentralization of R & D, and the extent to which R & D is internationalized. It includes the balancing of shorter-term, applied R & D and longer-term, more speculative basic research with potentially high rewards. It includes the management of creative and productive researchers and research teams.

MTI includes the closely related issue of the management of *new product development*, to which R & D is a major contributor. The management of new product development includes efficiency factors, encouraged, for example, through the use of various project management systems, and effectiveness factors, such as whether chosen new products complement and build upon firms' existing product bases, expertise, and reputations.

The management of *operations and production* includes a wide array of issues and has a highly specific literature. The focus here will be on strategic management with a concern for broader business and organizational issues

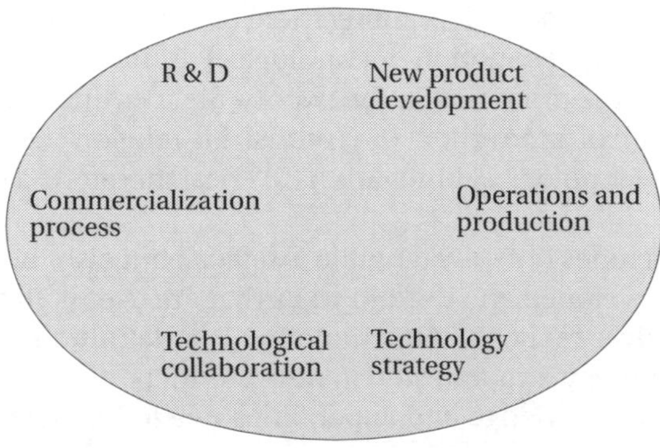

Fig. 1.1. Management of technological innovation: specific areas

rather than on decisions about hardware specifications. However, some of the major innovations in the management of operations and production, in, for example, quality, will be discussed. Specific focus will be placed on the move from mass to 'lean' production and the integration of manufacturing capacities with those of design and development, and the integration of supply chains.

Technology strategy is concerned with linking technology with firms' competitive strategies, and can be the basis of the overall strategy. It is concerned with integrating all the areas of MTI into a coherent whole. Technology strategy has as its focus the creation and use of particular competencies. These competencies are comprised of all the resources existing within a firm, and the innovative capabilities that enable it to change them. Technology-based small and medium-sized enterprises (SMEs) face particular issues in their technology strategy, especially in relation to the problems of managing growth.

Technological innovation rarely occurs through the activities of single firms. It is more commonly a result of inputs from a variety of firms, working together as customers and suppliers, or in various forms of alliances. MTI therefore includes *technological collaboration*, the management of technological alliances and networks.

The *commercialization process* is a central element of MTI. This involves consideration of intellectual property rights, licensing, the creation of technical standards, and the appropriability of firms' investments in technological innovation (the extent to which firms can ensure that they receive adequate returns from their investment). The commercialization process may not be immediate—firms have valuably expanded options for the future through the innovation process—and this has to be considered when evaluating its commercial returns.

The management of information technology (IT) is incorporated within these areas and is not included as a distinct area of MTI management. While there is a large, specific, and highly technical literature on the management of IT (Turban *et al.* 1996), many of the strategic and organizational issues discussed—the development of new IT products and software, and the use of IT in R & D and operations—are covered in the definition of MTI and are therefore discussed here.

Similarly, the management of technological innovation in services is not included as a separate issue. The boundary between services and manufacturing industry is increasingly blurred (Quinn 1992; Miles 1994; Gann and Salter 1998). Is a company that designs car engines a manufacturing firm or a service firm? Can IBM be considered a computer manufacturing firm when the focus of its strategy is to supply customer 'solutions' rather than products, and when its service division produces one-quarter of total revenue? Are software houses (sometimes known as software factories) using highly computerized writing tools making products? As Quinn (1992) argues, many important activities carried out by manufacturing firms—marketing, distribution, engineering, design, maintenance, accounting—would be counted as services if they were supplied externally. Many services, such as bank telling, are now delivered mechanically, and the value of manufactured products often lies in intangible attributes such as speedy delivery, convenience of use, brand identity, and reliability that, if not embodied in products, would be thought of as services (Lester 1998). Furthermore, many physical products are packaged together with intangible services. An example would be Ericsson's and Nokia's packaging of services around their products (Davies 1997). According to Quinn (1992), service activities contribute most of value-added in manufacturing and constitute 65–75 per cent of most manufacturers' costs. Rubenstein *et al.* (1994) argue that, as service companies create R & D capabilities—to deal with the need for increased product differentiation, reduce costs of developing and delivering services, and protect proprietary technology—they are dealing with many issues analogous to those in manufacturing firms. Therefore the majority of the issues discussed in respect to MTI can be applied equally to services and manufactures. Cases and examples will be provided of both manufacturing and service companies.

All areas of MTI can be managed for competitive advantage. For example, the ability of pharmaceutical and electronics firms to compete depends critically upon their capacity to manage R & D. It is R & D that provides the opportunities to create new products and markets. Drug companies such as Glaxo Wellcome and Merck rely on research to create highly profitable drugs for treating ulcers or producing anti-AIDS inhibitors. Companies such as Sony, with the Walkman, Samsung, with the DRAM semiconductor, BMW, with its Series 7 cars, Citibank, with its new services, and Netscape, with its web browser, depend on new products to provide the means by which they compete, and these new products to a large extent define their companies. Operations and production prowess gives companies like Toyota the ability to produce cars better and cheaper than competitors; allows US minimill steel producers, like Chaparral Steel, to compete effectively; and enables electronics companies, like Acer in Taiwan, to produce efficiently for major electronics-industry customers in the United States and Japan.

When NEC decided it wanted to develop expertise in semiconductors, which it saw as a key strategic technology central to its competitiveness in a number of industries, it used over 100 technological alliances to do so. The Italian computer company, Olivetti, claims it operates over 200 such alliances as a means of technology development and commercialization (Dodgson 1993*a*). Boeing is also heavily reliant on technological collaboration. It now works collaboratively in the production of aircraft, and partners are responsible for the design and manufacture of major components, such as engines, fuselage, and rudders. Boeing can no longer design and manufacture aircraft by itself.

Corporate history is littered with examples of companies that failed to commercialize their technological innovations. Ampex failed to see the real market for their developments in video recorders. RCA, famously, did not make the business transition from vacuum tubes to transistors. The effectiveness and qualities of the commercialization process determine the outcome of technological innovation. Sony's Betamax system was technically better than the competing VHS video system, and the IBM personal computer was in many ways inferior to other competing products. But the ability of Matsushita and IBM to commercialize and market their innovations provided their competitive advantage.

Of all the aspects of MTI, technology strategy is perhaps the most demanding. Very few companies have been capable of developing and implementing technology strategies consistently. Technological forecasting is notoriously difficult, but when it does occur significant competitive advantage can ensue (a good example is SAP, the German applications software company, which foresaw the importance of first using Unix then secondly Microsoft NT systems with its customers). Similarly, when market assessments are accurate and prescient—such as when Matsushita saw the market in home video recorders—market leaders can benefit significantly. Where firms have developed technology strategically, such as in the optoelectronics industry in Japan, or in flat-panel display technologies in Taiwan, they have established significant competitive advantages (Miyazaki 1995; Wong and Mathews 1998). Benetton, the Italian clothing company, has been particularly effective at integrating innovation in products, production, marketing, and sales to allow it to achieve its competitive aims of quickly delivering ever-changing fashion goods to market.

Some of the extensive empirical research findings showing the importance of technological innovation are listed below.

- High technology industries provide the highest rates of growth in productivity and employment in manufacturing industry (OECD 1996).
- Trade in high-technology goods (requiring high levels of R & D) increased from 11 per cent of international trade in 1976 to 22 per cent in 1996 (World Bank 1998).
- High-technology industries in the United States nearly doubled from $215 billion in 1988 to $420 billion in 1996 (chain weighted in 1996 dollars) (*Business Week*, 31 Mar. 1997).[1]
- Technological activities—as measured by R & D and international patenting—are statistically significant determinants of export and productivity performance (Fagerberg 1987).
- The returns to R & D investment, both 'social' (to society as a whole), and 'private' (to the firm making the investment), are consistently assessed to be high. In a study of seventeen innovations Mansfield *et al.* (1977) found the social returns of R & D investment to be 56 per cent, and private returns to be 25 per cent.
- Technological innovation has played a significant role in the economic transformation of the East Asian economies (Hobday 1995).

[1] All financial data in the book are denominated in US dollars.

- Entire industries, such as the Swiss watch industry, and geographical regions, such as Silicon Valley in California, can be invigorated or depressed by technological change (Saxenian, 1994; Utterback, 1994).
- At the corporate level, new products less than five years old account, according to one estimate, for 52 per cent of sales and 46 per cent of profits of US firms (Cooper 1993).
- Within the factory, the use of advanced manufacturing technology is unequivocally associated with greater productivity, higher survival rates, higher wages, and more rapid employment growth (US Department of Commerce 1994).

A brief note on new growth theory

Empirical findings on the significance of technology, are reinforced by new theoretical approaches that reveal the importance of technology, particularly new, or endogenous, growth theory. Traditional, neoclassical economics considers technology to be an 'exogenous' factor in explaining economic growth: essentially it is taken as given. Simply put, this form of analysis believes that productivity and growth are a function of combinations of the three productive factors: land, labour, and capital, with a large unexplained residual in the calculations. In this body of theory, technology may be part of the explanation for this residual, but there is little concern to establish its importance. The sources of technology and the distinctive and idiosyncratic ways technology is used in individual firms to create growth are ignored. Furthermore, technological investments, like all capital investments, produce declining returns over time.

In contrast, new growth theory argues that technology is an important 'endogenous' factor explaining growth, and comprehension of the way technology flows between firms and industries is essential (Romer 1990). Additionally, unlike conventional investments in plant and equipment, which generally have declining returns over time, technological investments are argued to produce positive returns (Arthur 1990). Both empirical analysis and theory show, therefore, that competitiveness and the ability to pay your way in the world depends on technological innovation.

The major features of new growth theory are listed below.

- *Technology is 'endogenous'—a central part of the economic system, a key factor of production along with capital and labour.*
- *Although any given technological breakthrough may appear random, technology overall increases in proportion to the resources devoted to it.*
- *Technology produces 'positive returns'. Traditional theory predicts diminishing returns to investment, yet sustained, robust growth can be achieved by technological investment.*
- *Investment can make technology more valuable and technology can*

make investment more valuable—a virtuous circle that can raise an economy's growth rate permanently.

- *Monopoly power is useful in providing incentives to technological research.*
- *The emerging world economy is based on ideas rather than objects and this requires different institutional arrangements and pricing systems taking into account, for example, that prices depend on development time, cost, and risk, not unit production costs.*
- *The possibilities for discovery and continual improvements are endless.*

GENERAL ISSUES TO BE MANAGED

A consideration of the specific issues of MTI has indicated its significance to business. It is difficult, however, to think of a more challenging aspect of contemporary management. Although competitive advantages can be derived from the good management of the specific areas described above, managing technological innovation involves a number of general issues affecting technological competitiveness.

MTI involves trying to manage something which is inherently *complex* and *risky*. In addition to the intrinsic complications of many products, a key aspect of complexity lies in the systemic nature of contemporary industrial production. Technology-based innovations, be they aeroplanes, cars, buildings, home banking, or personal stereos, are comprised of various component systems. Computers, for example, comprise central processing units, operating systems, applications software, disk drives, memory chips, power supplies, and communications devices. Integrated Services Digital Networks (ISDN) involve the participation of telephone companies, satellite suppliers, microwave vendors, local area network companies, and value-added network operators. The integration of these often highly complex systems is a key MTI task.

Some of these complex systems have been described as a different form of industrial production, requiring different management approaches (Hobday 1998). Thus, for complex products and systems (CoPS), including high value products, capital goods, control systems, networks, and civil engineering constructs such as aircraft engines, avionic systems, offshore oil equipment, and intelligent buildings, the emphasis is on design, project management, systems engineering, and systems integration. According to one estimate, these complex systems accounted for about 11 per cent of the United Kingdom's GDP in 1994 (Gann and Salter 1998).

Risk is determined by a number of considerations. Important factors in the management of risk in technological innovation include unpredictability, cost, and appropriability. The innovative activities of firms, for example, are confronted by general *business* uncertainty of future decisions on investment; *technical* uncertainty about future technological developments and the parameters of technological performance and cost; and *market* uncertainty about the commercial viability of particular new products or processes (Freeman and Soete 1997). With the high degree of risk of investments in technological innovation (see Chapter 3), and the very high levels of investment in it (some firms spend billions of dollars annually on R & D, and some industrial sectors, such as electronics and pharmaceuticals, spend over 10 per cent of sales on R & D), there are enormous pressures internationally to reduce the costs of technological innovation. One method of reducing uncertainty in investments in manufacturing plant and equipment is investment in flexible technologies. Chapter 5 will argue how such flexibility allows firms successfully to adapt to changing market circumstances.

Firms also need to manage the appropriation of returns to their innovations. There are risks associated with the methods used to ensure desired returns to R & D investment, such as whether intellectual-property protection is awardable and can be maintained. An additional consideration concerning appropriability is the question of speed: how quickly can innovation be protected and returns achieved? New markets can develop very rapidly on the basis of new technology. In the decade since its development, it is estimated that global electronic-commerce on the Internet will, by 2003–5, have become a 1 trillion dollar business (OECD 1998).

Whether applied to developing or improving new products and services, or producing existing ones, corporate competitiveness requires *knowledge*, and the *organizational ability to learn fast* and move quickly when winning notions emerge. All of the issues of MTI depicted in Fig. 1.2—complexity, creativity, risk, knowledge, and learning—will be examined throughout the book.

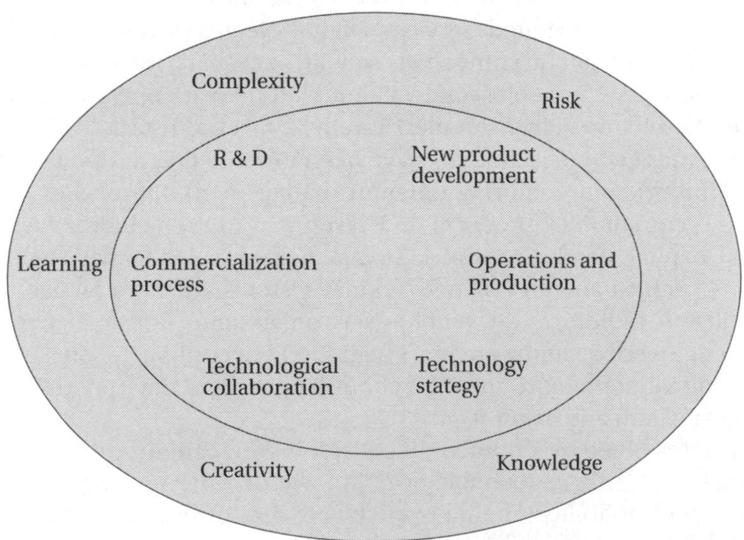

Fig. 1.2. Management of technological innovation: general issues

CASE STUDIES IN TECHNOLOGY INNOVATION MANAGEMENT

Some of the aspects, issues, and common problems of technology and innovation management will be briefly illustrated in the following short case studies. They are composite descriptions of actual companies and highlight the problems they face in MTI.[2]

The British pump firm

A British pump firm provides us with a classic example of a traditional manufacturing company, and the changes confronting its managers. Some British pump firms have been operating for over 100 years, but few have successfully diversified, or grown to any significant size, and the industry has seen numerous company closures over the past decade.

The managing director of the pump firm might remember that in the 1970s and 1980s innovation would have been driven by local sales representatives reporting customer dissatisfaction with a particular facet of the product. Occasionally, innovation might have been driven by the R & D or engineering function of the firm finding some new option that could prove attractive to purchasers.

Today, the same managing director is confronted by a radically different and more challenging global competitive environment where innovation is the key to survival and growth. Cheap, high-quality pumps are available from lower-wage economies. Exceptionally efficient pumps using lightweight materials and less energy are offered from a wide range of advanced economies, all with distributors or licensees in the United Kingdom. In order to remain competitive or improve its competitive position, the pump firm has to be exceptionally smart in its innovation activities so as to deal with the increasing complexity of its products. It must be intimately aware of customer needs; indeed, it may occasionally have to anticipate them. It has to be abreast of recent scientific and technological advances in new materials and designs, and it requires links with research groups in universities and research institutes taking on problem areas in pump technology, like cavitation, which it cannot solve itself.

[2] The description of the pump firm derives from a research project in the British pump industry conducted by the author, Professor Ron Dore, and Dr Hugh Whittaker. The Japanese electronics firm case is based on a study of the Japanese multimedia industry conducted with Professor Mari Sako, and research into a major Japanese electronics company in Singapore. The biotechnology example was based on the author's study of Celltech and the biotechnology industry. The Taiwanese case is based on numerous research visits to Taiwan, encompassing both large and small firms, and the Industrial Technology Research Institute. The author is an adviser to the Taiwanese National Science Council's research programme in the machine tool industry. The Indian software company was based on a number of research visits to India and particularly to Bangalore, and on a case study of the international technology strategy of Ericsson. Some readers may notice that male managers predominate in these case studies. This gender imbalance reflects the actual situations of the large number of firms studied.

The company has invested in an expensive computer-aided design (CAD) system to assist its design processes. This system is being integrated with its production system to facilitate the efficient manufacture of its increasingly complex designs to the high levels of quality expected in the industry. Through this electronic integration it is now designing parts that it knows can be made. In the past occasions arose where substantial development work went into a product that subsequently could not be manufactured as the cost of retooling the factory would have been prohibitively expensive. The managing director is sensitive to the need for the company's internal organizational processes to be highly efficient so that creative ideas are acted upon quickly. The different functions—marketing, engineering, production, distribution, and service—are integrated, and new product development teams are well managed, highly skilled, empowered, committed, and have the necessary funds available to undertake investments in more unpredictable markets.

The market has changed considerably. Pumps are often components of complex systems—in buildings, factories, power stations, sewerage works. Customers require the firm to learn to integrate its products into the various systems of assemblers or contractors, sometimes overseas and often operating to different technical standards. The pump firm must sustain continual dialogue with the firms coordinating the creation of the systems (the systems integrators) to produce creative solutions. Great efforts are being made to integrate its CAD system with them through sharing common databases and design practices. It is also developing closer links with its suppliers in, for example, electronic control systems, so that it is aware of any advantages to its customers that might derive from their developments. Rather than being the passive instrument of the systems integrators, whose control over the system ensures them substantial profits from their negotiating strengths, the firm has ambitions to become a systems integrator itself. It sees its investment in CAD as providing the opportunity for it to develop these skills in designing and integrating new product systems.

The company directors operating in this environment are aware that all these activities must be driven by a clear strategy such that all efforts fit together in cumulatively developing a distinctive advantage over competitors in technology, and establishing a clear reputation both in the marketplace and as an employer attractive to highly skilled, creative, and committed workers. They know an international approach is needed towards both market and technology, as many customers and suppliers have overseas operations.

The US biotechnology firm

The biotechnology firm is a relatively new phenomenon that emerged in the late 1970s in the United States. These firms began as vehicles for transferring new scientific discoveries in genetic engineering and immunology into industry from government research laboratories and universities. Some firms were initially expected to follow the pattern of the IT industry and duplicate the remarkable growth of firms such as Apple, DEC, Intel, and Microsoft. Few biotechnology firms, however, have grown to any size, and most remain focused on product development, rather than on becoming integrated producers and distributors of products such as pharmaceuticals.

The MTI challenges facing the biotechnology firm are considerable. Our firm was started by two scientists and a venture capitalist on the basis of a scientific discovery with two potential market applications in health care. Laboratory tests had proven very successful, and the scientists believed their discovery could assist in overcoming ailments with a market of several hundred million dollars in the United States. The venture capitalist had achieved considerable success in the computer industry, but had little knowledge of the drug business.

A major challenge confronting the biotech company is the regulatory process needed first to protect, and second to develop its discovery. The company had patented its discovery (which was the basis of the venture capitalist's investment). There were, however, a number of technical aspects allied to the major discovery that had not been fully patented. This resulted from a certain naïvety on the part of the directors of the new company, and a concern to control the costs of patent registrations. Subsequently, the company had discovered that the real commercial value-added of its discovery was not the substance itself (a complex protein) but the process of scaling up and manufacturing the product. This process is delicate, involving growing the product in quantities of a few grams, using the medium of a specific animal gene. Considerable intellectual capital had been invested in mastering this production process, but it had not been patented, and competitor firms had also mastered this technology because the two scientists had continued their academic tradition of publishing and discussing their research findings. Although the company knew it was in the knowledge-selling business, it had not realized which aspect of its knowledge was most valuable.

A second problem facing the company was the amount of time and money it takes to gain approval for the development of a new pharmaceutical. In the regulatory context of the United States, it can take between four and fourteen years, and cost in excess of $250 million, to secure approval for a new drug, as all drugs go through a strictly controlled testing and approval process. The company had initially focused on developing one of the two potential applications. It had found, however, that, although the new product worked, it did not perform demonstrably better than existing products on the marketplace. So the other application was being pursued, and this had involved considerable delay and increased cost.

Early in the process it had become apparent to the company that it could not afford to proceed through the regulatory process of drug approval by itself, nor could it possibly develop the huge marketing and distribution effort required to bring its products to market. It had initially targeted an over-the-counter product available from pharmacies, but had subsequently decided that the cost of commercializing such a product would be prohibitively expensive. Instead it explored the possibility of targeting hospitals, but this also proved too expensive. To improve its cash flow, it had begun offering research services to other companies, using its expertise and scientific equipment to analyse and sequence various genetic materials. It had, after much debate and with some reticence, also entered into a strategic alliance with a major US pharmaceutical company, receiving substantial investment capital in return for rights to the developed product.

The management challenges facing the biotechnology firm's three directors are therefore considerable. The two scientists, rather than doing research, find their time consumed with liaising with regulatory bodies (to deal with patent infringements and the drug-approval process of the US Food and Drug Administration), performing routine procedures to assist cash flow, and managing the sometimes difficult and demanding relationship with the large pharmaceutical firm. How do the scientists maintain the aura of excitement and discovery in the company, and encourage the creativity required for continuing new product development and building the firm's knowledge base? The venture capitalist, whose expectations of fast returns have not been fulfilled, has to decide about her exit strategy. Does she continue to bankroll the company until its product is developed or nearly developed and then sell the company in an Initial Public Offering, potentially making very substantial returns? Or does she continue to sell the company's intellectual property at a much lower return to a larger pharmaceutical firm and expose herself to less risk?

The company faces important strategic decisions about its future. Does it become a research services company (and where is the fun for the scientists in that)? Does it continue to fund its own expensive research in order to develop a pipeline of new products? Does it become very ambitious and attempt to develop and market its own products, perhaps in collaboration with other firms?

The Taiwanese machine tool company

Machine tools are a key technology in manufacturing industry. They make the components of all machines, including, of course, the parts of which they themselves are made. They perform the various processes of cutting metals and other materials, such as milling, turning, drilling, and boring. These processes are heavily computerized, ensuring high quality and standardization. They are also increasingly complex in the extent to which they are integrated into other aspects of manufacturing. In planning, factories operate using shared databases controlling production sequences and linking design and production. In production, machine tools are linked with robots and automated transfer mechanisms in flexible manufacturing systems. Taiwan has a thriving machine tool industry, encouraged by supportive government policies.

Our machine tool company began by producing traditional, non-computerized machine tools for turning simple components for the domestic bicycle industry. The founder of the company sent his son to study engineering in the United States. After staying on to complete his MBA and to work for a few years for a supplier to the US aerospace industry, the son returned to Taiwan and immediately set about upgrading the company's technology and product range. He invested heavily in new design technologies and databases and embarked on a strong export drive assisted by favourable loans from the government. He introduced computer control of the machine tools by purchasing components from a Japanese supplier. He improved the design and functionality of the products (so that they could perform several machining functions) through working closely with a government research agency. The research agency had been funded by the Taiwanese government to develop the machine tool industry and had assiduously collected information about technological advances around the world. It had created a research group in Taiwan that undertook collaborative research projects for machine tool producers, and our machine tool company participated in and helped direct the technical aims of a number of these. The company had recruited back to Taiwan a number of first-class researchers and engineers working in the United States to help with these developments. Its recent investments have been in advanced product management systems and software that enable it easily to store and retrieve design data, prototype electronically, and allow closer integration with its suppliers.

The company faces a number of problems. First, the imported Japanese computer/software component of the product is becoming an increasingly important element of the product's cost. A strong yen, and an inability to control supply, make the company wish to produce the computer controls domestically. Should it do so itself, in an area where it has little expertise? Should it acquire one of the many innovative local computer companies and focus it on machine tool controls? Or should it collaborate with a local computer company?

Secondly, competition has become intense in the standard machine tool market. The demand in export markets and in Taiwan lies in highly sophisticated machines capable of cutting new materials to extraordinarily high precision, in production contexts ranging from the aerospace industry to the increasingly sophisticated bicycle industry. The company needs to develop expertise at the forefront of the interfaces between mechanical and electronic engineering. It requires basic scientific knowledge about complex mathematical calculus and the properties of new materials. The company has been superb at catching up with world best practice and is now at the technological forefront. Its future competitiveness depends on developing technological leadership and managing the substantial risk this involves. In addition, the research institute, although helpful in the past, has become increasingly less capable of assisting a firm on the technological frontier. The system of government support would need to change radically in order to continue to assist this and similar firms.

The Japanese electronics firm: the corporate R & D laboratory

The Japanese electronics industry is one of the industrial success stories of the second half of the twentieth century. Throughout its development, from catching up with industrialized economies after the Second World War to its present position of international technological leadership in many fields, its firms have engaged in substantial R & D.

The large parent company of our corporate R & D laboratory is one of the world's most successful consumer electronics companies. It spends nearly $1 billion on R & D each year, mostly on the company's fifteen decentralized divisional laboratories. Around 10 per cent of the company's total R & D spending is allocated to the central laboratory, which develops longer-term research (defined as having expected outcomes beyond five years). This laboratory has been successful in providing scientific support to the company's divisional research functions, and its researchers are highly productive, as measured by the number of academic publications and patents produced. It has successfully added to the technological options available to the firm.

The research director of the central laboratory is facing a number of conflicting pressures. He has to extend the range of expertise within his laboratory to meet the technological requirements of the divisions and firms he supports. At the same time, because of adverse macroeconomic circumstances, his budget is being reduced and he is under strong pressure to speed up the returns of research efforts. The core areas of science underpinning the company's activities are becoming broader and therefore less controllable, and he no longer has the breadth of knowledge in his staff, or the range of scientific equipment, required to undertake the research he considers necessary. He fully understands the reasons for the firm's desire for faster results from its R & D investments, but knows that his lab's major contributions to the company in the past have been through longer-term, more basic R & D.

The expertise required by the firm ranges from abstract theoretical particle physics to the development of new generations of embedded software. The company has overseas research laboratories linked to universities with particularly advanced expertise in these areas. While this system is working well in searching for and bringing excellent information back to Japan (assisted by Internet communications), he is concerned about maintaining sufficient levels of expertise within his organization to be receptive to the wide range of inputs he requires. The research director is also finding it difficult to manage the international R & D labs, particularly the task of converging their mission of undertaking basic research with the greater demands he is facing for quicker returns to the firm's investment. He is also under some pressure from within his company and from his contacts in the Japanese government to increase the amount of work with local universities, whose scientific expertise in needed areas is gradually increasing but is still some way behind that found elsewhere.

One opportunity the research director sees is in collaborative and subcontracted R & D. The central lab has been engaged in a number of Japanese government-sponsored collaborative R & D projects, with varying degrees of success. While his staff have been uncomfortable sharing scientific research with companies with which they vigorously compete, experiences of working collaboratively with smaller overseas firms, which are generally much quicker at commercializing basic research, have been highly positive. The research director is also aware of the potential of a number of highly specialized local small research and software firms, whose creativity he wishes to access. He is aware, however, of the dangers of imposing large-firm management controls and reducing the distinctive advantages brought about by their flexible, unbureaucratic structures and incentive systems.

The Indian software company

Based around several major cities, such as Bangalore and Hyderabad, the Indian software industry has grown since the 1980s to be an international leader in software production. Indian software engineers are renowned for their high technical skills and comparatively low labour costs. It is these attributes that have attracted companies such as Microsoft, IBM, and Intel to make substantial investments in India.

Our software company was started by two brothers, both graduates of the prestigious Indian Institute of Technology. One brother had worked as a programmer for a large German electonics company in India, and the other had worked for a number of years as a software engineer for a major US software company. They inherited some family money and decided to start a company together. The company recently recruited a chief executive, who was an experienced manager from a Canadian telecommunications company (and a family relative).

The company is based in Bangalore. The decision to locate there was made primarily on the basis of its large labour market for programmers and software engineers. Many multinational IT companies had also located there. Furthermore, and importantly, with less pollution and traffic congestion, the general working environment was more agreeable in Bangalore than in many Indian cities. The city is famous for its nightlife and bars—an important drawcard for the young workforce.

The company is ten years old and employs forty-five people, only three of whom (including the oldest brother and the chief executive officer (CEO)) are over 35. It has been working as a 'software factory', writing millions of lines of code as a systems software subcontractor, mainly for US software companies and for two other locally based companies. Orders come in with highly specified requirements, and the company writes the code on a jobbing basis, mainly using software tools, such as computer-aided software engineering. One of the attractions of the company to its clients is its strict adhesion to quality management. As a legacy of the founding brother's association with the German company, a large amount of time and resources have been dedicated to winning ISO 9000 approval. The company is one of the few in Bangalore to hold all relevant ISO quality management approvals.

At first sight, the company is doing well. It is profitable and has good relationships with its customers. However, the two brothers are concerned about the future, and these concerns led them to appoint a new CEO. One of the major problems confronting the company is a tightening labour market. Salaries are increasing rapidly, and good software engineers are now in a position to pick and choose between employers. Whereas a few years ago salary level was the primary consideration for employees, more recently the company has been losing employees (and staff it wanted to recruit) to other companies offering more interesting and varied work and greater leisure time. The profit margins of the company are becoming squeezed, and the company's largest client has opened a software company in China, where salary levels are significantly lower.

A second problem lies in the rapid growth of the company. Neither of the company founders had managerial experience or training. Although the business had succeeded in the past by means of its excellent project and quality management skills, it had 'flown by the seat of its pants' in other areas, such as marketing and human resource management.

The new CEO is aware of these challenges and has a number of strategic options. He believes that software subcontracting of the sort the company has done in the past is a 'race to the bottom'. Prices and margins will continue to be squeezed. He knows that competitiveness depends on attracting the best and brightest workforce, and wants the company to be attractive to the most talented and creative local employees. His business strategy has two elements. First, he plans to start selling applications software services to companies in the telecommunications industry, a sector he knows well. The company has experience in working in open, client-server, architectures and feels it can produce software that fits seemlessly with customers' current environments. Secondly, he realizes that the company's project management skills give the firm the opportunity to coordinate other local subcontractors as a 'systems integrator'. Here particularly the company's edge derives from expertise in using the capability maturity module (CMM), a package developed by the National Aeronautics and Space Administration (NASA) to measure how effective a software organization is in managing software projects to budget, and its ISO quality approvals. In effect, he plans to position the firm as the prime contractor, which will coordinate a network of further subcontractors. He is planning a reorganization of the workforce and is planning to promote five of the most high-potential engineers to project management positions.

His primary challenges include developing a technologically aware marketing function. Previously, the company's marketing was reactive and domestic in nature. It now needs to become international and proactive. He needs to develop the management skills of the workforce such that it can manage both clients and its own new software products, and at the same time develop the 'network management' skills required to coordinate local software suppliers. The company also needs to continue to develop its effective use of tools like CMM as it expands its technological ambitions.

CONCLUSIONS

This chapter has defined MTI, analysed its importance, and illustrated some of its key elements. It shows why MTI is a major contributor to the construction and maintenance of competitive advantage. Whether considered at a national level, or in a particular industry or firm, technological innovation is important, and its effective management is crucial.

MTI includes: the management of R & D, new product development, operations and production, technology strategy, technological collaboration, and the commercialization process. These activities are often complex, involving high levels of technological and organizational integration, and risky, including a high level of unpredictability, concern to control costs, and manage appropriability. In all considerations of technological innovation the important issues of knowledge and learning are critical, and these will provide major underlying themes of this book.

As our examples have shown, and as will become clear through many other examples in this book, MTI is a difficult process, and even companies that have been highly successful at it in the past are confronted by substantial new challenges. These challenges range from the day-to-day operational issues of how to make CAD systems more efficient, or dealing with government bureaucracy in order to meet regulatory requirements, to major strategic issues determining the future of the company. The British pump firm needs to decide whether it wishes to, and can, become a systems integrator; the US biotech firm needs to decide whether it becomes a research services company or an integrated pharmaceutical company or something in between; the Taiwanese firm has to decide how it is going to access a key technology for the future, either through acquisition or through collaboration; the Japanese laboratory needs to reconcile its strategic, long-term function with short-term financial constraints; and the Indian software firm needs to decide how it is going to manage its future growth by producing more creative, higher value-added services. Many of these challenges result from broader changes occurring in industry and business and it is to those challenges that we now turn.

CHAPTER

2

The New Challenges of the Management of Technological Innovation

Some of the challenges confronting firms in their management of technological innovation were described in Chapter 1. These challenges are reflections of major changes in the nature of industry and business. The most significant of these changes will be examined in five contexts: in industry in general, in business and innovation systems, in management, in the innovation process, and in globalization. To comprehend the future challenges of MTI it is necessary to understand these broader contextual and environmental changes.

THE CHANGING NATURE OF INDUSTRY

The US car worker of the 1930s might have some basic recognition of the way cars are produced at the beginning of the twenty-first century at Fiat in Turin or Nissan in Sunderland (and the prevalence of robots might confirm some long-held suspicions), but would have little comprehension of industry as we know it. If he were transported to a contemporary US car plant, he would be amazed to learn that there cars were all designed, and some major components tested, by computers in electronic offices. The range of new materials being used in the car, its instrumentation, the efficiency of its engines, and the range of safety designs and aids offered would probably confound him. He would almost certainly be surprised to learn that the car body was designed in Germany, its engine in the United Kingdom, while components were made in Mexico, Spain, and Australia. Most perplexing of all, no doubt, would be his discovery that he was working for a firm whose organizational and production practices were essentially Japanese.

As industry continually changes and, in some cases, can alter radically in the course of one generation, it is necessary for managers to understand the historical forces shaping its organization and conduct. One of the forces with a significant impact on industrial development is technology, and an in-depth analysis of technological change is vital for effective management.

Technological waves of development

In the 1930s the German economist Schumpeter noted that technological innovations are not evenly distributed over time or across industries, but appear in periodic clusters. Since the Industrial Revolution, it has been possible to identify historical waves of intense technological change ('technological revolutions') characterized by rapid economic growth opportunities and radical social changes (Freeman and Perez 1988). These revolutions (which are described as changes in *techno-economic paradigm* by Freeman and Perez) depend upon clusters of mutually supporting technological innovations being accompanied by social innovations in areas ranging from organization and management to taxation and employment law. Fig. 2.1 illustrates these historical waves in a simplified form and describes the key 'factor industries' associated with each wave. Factor industries, such as cotton, steel, and oil, are typified by continually reduced costs, readily available supply, and an impact across wide areas of the economy.

According to this theory we are presently in the fifth wave of technological

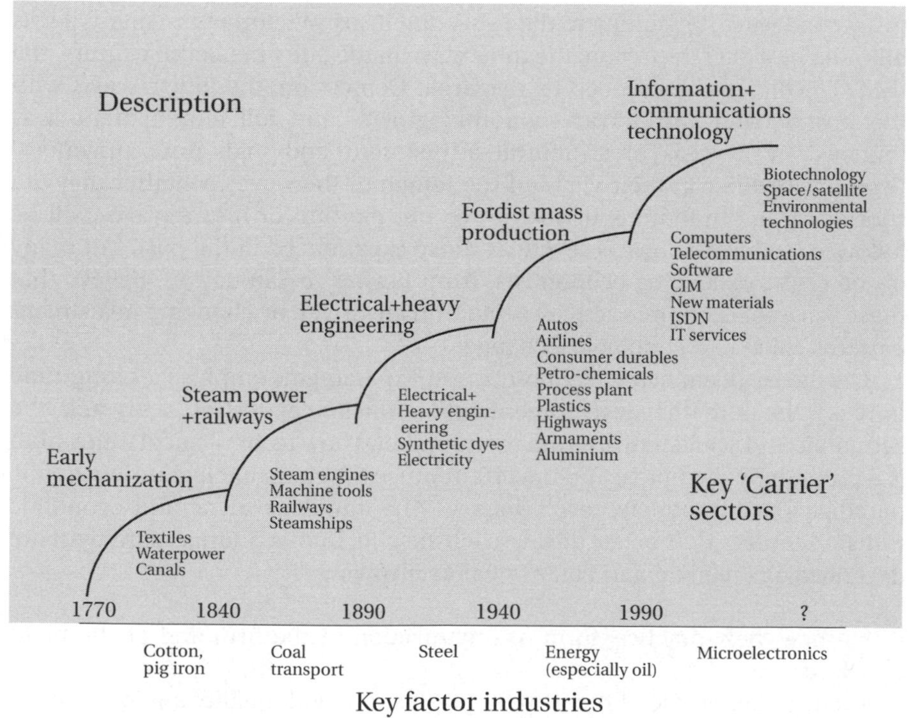

Fig. 2.1. Waves of technological development, 1770–1990

development, the information and communications technology wave, with microelectronics being the major factor industry. Soete (1991: 45) argues that major adjustments resulting from the fifth wave include a redesign and new configuration of the capital stock, a new skill profile in the labour force, new management structures and work organization, a new pattern of industrial relations, and a new pattern of institutional regulation at national and international levels—for example, in relation to the global telecommunications network or traded information services.

The theory of changing techno-economic paradigms may help explain why some economies grow faster than others. As Freeman (1994: 88) puts it, those nations that prove most adept in making institutional innovations that match the emerging new techno-economic paradigm are likely to prove the most successful in growing fast, catching up, or forging ahead. Those, on the other hand, that suffer from institutional 'drag' or inertia may experience a prolonged mismatch between their institutions (including management systems at firm level as well as government structures), and the growth potential of new technologies.

As the concept of 'waves' of economic activity implies, periods of economic growth are followed by recession and depression. Thus, the first wave, the Industrial Revolution in the late eighteenth century, was followed by recession; the second wave, beginning in the 1830s and 1840s ('Victorian prosperity') was followed by a deep recession; the third wave in the late nineteenth century, the '*belle époque*', was succeeded by the Great Depression; the fourth wave, with the post-Second World War economic growth and full employment, was followed by a crisis of structural adjustment and high unemployment. Economists differ in their views of the length of the waves, whether they are shortening in duration, and whether or not the fifth or future waves will be followed by downswings as severe as those experienced in the past, but many of the century's leading economists, from Keynes to Samuelson, believe that these waves of economic activity occur and are driven by changing investment patterns allied to technological change.

The theory of changing techno-economic paradigms emphasizes long-time horizons in both the development and diffusion of technologies and the economic and social returns from them. It shows how technological innovation is a profoundly disruptive and uncertain process—changing techno-economic paradigms are genuinely 'revolutionary'. The immense social and economic transformations that occur during each new techno-economic paradigm are described by Freeman and Perez (1988) as involving:

- a new 'best-practice' form of organization in the firm and at the plant level;
- a new skill profile of the workforce, affecting both quality and quantity of labour and corresponding patterns of income distribution;
- a new product mix, with new technologies representing a growing proportion of gross national product;

- new trends in innovation (both incremental and radical) as substitution of the new factors occurs;
- a new pattern of location of investment both nationally and internationally as the new factors change relative comparative advantages;
- a new wave of infrastructural investment to encourage diffusion of the new technologies;
- a new wave of entrepreneurship and small, start-up firms in new technologies and industries;
- a tendency for large firms to concentrate—by means of growth or diversification—in the new factors;
- a new pattern of consumption of goods and services and new types of distribution and consumer behaviour.

Just as these transformations have occurred in the past, so they are strongly in evidence at the turn of the century.

Box 2.1 outlines some of the key features of major industrialized nations found in the 1960s and 1970s and contrasts them with the present situation. All these changes have profound effects on the ways of managing technology and innovation. Many of the issues will be discussed in greater detail in subsequent chapters.

The significance of all these changes is summarized by Lester (1998: 322) as follows:

> During periods of rapid change, investment in intangible assets—knowledge, ideas, skills, organizational capabilities—takes on special importance. The results of these investments—ideas for new products and processes, knowledge of new market possibilities, more competent employees, nimbler organizations—give the economy the flexibility to keep adapting and reconfiguring itself to new supply and demand conditions. They are the lubricants of the economic machinery.

Just as significant is the likelihood that the extent of the changes that have occurred since the 1960s will be matched by similarly dramatic changes in the 2030s and 2040s. Many of the changes outlined have led to descriptions of late-twentieth-century industry and society as 'knowledge based'.

The knowledge economy

The World Bank (1998: 1) argues that 'knowledge has become perhaps the most important factor determining the standard of living—more than land, than tools, than labor. Today's most technologically advanced economies are truly knowledge-based'. The knowledge economy is not only about new creative industries and high-tech business, it is relevant to traditional manufacturing and services, and to businesses ranging from construction and engineering to retailing and banking (UK Department of Trade and Industry 1998). It is estimated that more than 50 per cent of gross domestic product (GDP) in the major OECD economies is knowledge based (Stevens 1996).

Box 2.1. ***Major features of industry, 1950s–1990s***

1950s and 1960s 'Convergence & aggregation' (the 4th wave?)	1990s onwards 'Divergence & disaggregation' (the 5th wave?)
Dominance of large-scale, vertically integrated firms	Decentralized, network-based, flexible firms
Mass production systems, dedicated machinery	Lean production systems, flexible machinery
Mass, stable, standardized markets	Niche, rapidly changing markets, customer sovereignty
Centralized management	Decentralized management
Monopoly and oligopoly	Intense competition
Strongly directive government, state-owned utilities and telecoms, protectionist industry policies, tri-partisanship between government, unions and employers	Non-interventionism, privatization and deregulation, government as regulator not provider, free-trade policies
Strong role of trade unions: from policy-making to demarcation decisions	Declining power of unions, employers' concern for 'employees', multiskilling
Separation of management and ownership	Share-owning incentives and management buy-outs
Full-time secure employment	Significant part-time, contractual employment
Some internationalization of industrial production	Globalization of business
Nationalism in trade and industry policies	Pan-nationalism in trade and industry (EU, NAFTA, APEC)
Predominance of Western models of management	Integration of international best practice in models of management
Science and research undertaken in universities and large firms	Substantial increase in scale and scope of science and research and diversity in provision ('the new production of knowledge')
Technology development a feature of individual firms; not-invented-here syndrome; anti-trust legislation	Technological collaboration a feature of government policies and corporate strategies

Clear distinction between manufacturing, services, and resources sectors	Blurred boundaries in the knowledge economy
Competitiveness derived from tangible assets: capital, land, and labour	Competitiveness derived from intangible assets: skills, capabilities, creativity.

There are a number of drivers behind the move to the knowledge economy. First, the increasing knowledge intensity of the processes of generating, producing, and commercializing new goods and services (these will be examined throughout this book). Secondly, the almost exponentially extended capacity of information and communications technologies to store, process, and transfer vast amounts of information. Thirdly, the process of globalization (examined in a later section).

According to the OECD (1999), investments in knowledge have grown slightly more rapidly than GDP in the OECD since the mid-1980s. Such investments—which include spending on R & D and on software, and public spending on education—represent 8 per cent of OECD-wide GDP, a level similar to investments in equipment. If privately funded education and training is included, the figure exceeds 10 per cent.

Indicators that can be used to illustrate the growing 'knowledge intensity' of developed economies include their increasing expenditures on R & D and information and computer technologies, and the increasing proportions of high-tech goods in world trade in manufactures (Sheehan and Tikhomirova 1998). Johnston (1998) argues that the knowledge intensity of trade has increased steadily and persistently, from an index value of 0.70 in 1977 to just over 1 in 1994, rising by 44 per cent in this period. (The index of knowledge composition of manufactured exports is derived by weighting the exports by industry of a given country by the average knowledge intensity of the industry in the major OECD nations.) Measuring the market value of a company to its book value (Tobin's q) provides some indication of the intangible assets or knowledge in a company. The q value of the 1996 stock market was 1.3 in the United Kingdom, and 1.5 for the United States. Knowledge-based companies, like Zeneca in the United Kingdom, measured 5, while US-based Microsoft registered at 15 (UK Department of Trade and Industry 1998).

In addition to the changes in the extent of the importance of knowledge, the ways of producing new knowledge are also changing. Gibbons *et al.* (1995) argue that increased competition in industry has led to increased supply and demand in a market for knowledge. They distinguish between 'traditional' and 'new' modes of producing knowledge. Alongside traditional, disciplinary knowledge (which they call Mode 1), a new, broader, transdisciplinary and highly contextual form of knowledge is emerging (Mode 2) and supplementing Mode 1 (Box 2.2).

Box 2.2. ***Modes of producing knowledge***

Mode 1	Mode 2
Problems set and solved in a context governed by academic community	Knowledge created in the context of application
Disciplinary	Transdisciplinary
Homogeneity of producers	Heterogeneity of producers; encouraged by information and communications technology
Hierarchical and continuing	Hetarchical and transient
Quality control through peer review	Socially and economically accountable and reflexive quality control
Emphasis on individual creativity	Creativity a group phenomenon

According to this thesis, knowledge production occurs in an increasing variety of organizations, in firms and consultancies as well as traditional universities, and in an increasing number of ways. Consequently, the form of the links between knowledge-producing organizations becomes important (and may follow patterns different from those in the past).

It is the change in techno-economic paradigm created by biotechnology that has provided business opportunities for start-up firms like the one in the case study in Chapter 1. The United States has led the world in the commercial development of biotechnology, and its universities and government research laboratories have undertaken the basic research to create the seed corn for new businesses, many of which have emerged in a new organizational form: the biotechnology firm. Economies that are adaptive and have the capacity to encourage new industries and entrepreneurship reap the advantages from the new technology. Government policies for intellectual-property protection, and regulations that facilitate rapid and safe drug development, have encouraged the development of the technology. A large and highly technology-conscious venture capital industry has provided another form of institutional support. Industry has adopted new forms of organization, such as the close relationships between large pharmaceutical firms and small biotechnology firms.

Our entrepreneurial biotechnology firm was created by a new wave of technological development. Its operations are encouraged by a supportive national innovation system. Its growth relies on new sources of finance, organization, and management. And its business future depends less on tangible products than on intangible knowledge, protected by patents. Although its original knowledge base was derived from traditional, if interdisciplinary, research in a university, much of its present knowledge is self-created, resulting from its leading edge R & D. The sources of knowledge production in this new technology have been dispersed and expanded.

BUSINESS AND INNOVATION SYSTEMS AND NETWORKS

Analysis of the ways contemporary business is conducted and innovation occurs has led to an appreciation of the importance of various forms of *system* combining the activities of firms and institutions. Rather than being autarkic, atomistic organizations, firms are part of numerous systems and the form of these has many implications for MTI. Recent studies have identified the range of actors and institutions involved in business and innovation systems and networks, the way these combine competitively and cooperatively, and the social and cultural basis of these interactions (Edquist 1997). By highlighting the systemic nature of business and innovation, they can usefully identify weaknesses that can then be addressed to enhance systemic strength.

Business and innovation systems and networks are themselves changing with the international transfer of good practice, seen, for example, in the transfer of Japanese production and new product development methods to the West, and US R & D management practices to countries like Japan, Korea, and Taiwan. Nevertheless, there remain sufficiently important characteristics of these systems that are nation specific—in social, cultural, and institutional terms—to support a form of analysis known as national innovation systems.

National innovation systems

The system of innovation that has received most analytical attention is a country's *national innovation system* (NIS). Recognition of the systemic nature of innovation has led, during the 1990s, to international work on NIS as part of the search for the sources of national competitive advantage. Studies of NIS began with Freeman's (1987) analysis of the Japanese system. Since then, the NIS approach has been used in a number of country studies, most successfully in analysing the phenomenal growth of the Korean economy to the late 1990s (Kim 1997).

There are a number of analytical approaches to NIS. Without oversimplifying, one of these approaches can be described as the 'institutional approach'. It examines the relationships between the national institutions of finance, education, law, science and technology, corporate activities (particularly those that are research oriented) and government policies, and their influence on the propensity for innovation (Nelson 1993). Another form of analysis can be described as the 'relational approach'. This analyses the nature of business and social relationships in nations, manifested, for example, in the way links between suppliers and users of technology encourage shared learning (Lundvall 1992). This approach focuses on the importance of socially embedded knowledge and learning. Patel and Pavitt (1994: 79) define national innovation systems as 'the national institutions, their incentive structures and their competencies, that determine the rate and direction of technological learning (or the volume and composition of change-generating activities) in a country'. Included in this approach is the role of incentives such as temporary monopoly profits to encourage basic research. Also included are the strategic capabilities of firms, whose differences have considerable impact upon technological innovation and competitiveness.

Although studies of NIS correctly point to the activities of firms as the key determinant of success, they also highlight the importance of government. This means that MTI managers need to be acutely aware of the regulatory systems in which they operate, particularly in respect of intellectual property rights, standards and environmental issues, and the ways these are changing. They also need to understand government procurement and outsourcing policies. Many governments around the world are looking for increased contributions to national R & D efforts from the private sector, and for faster financial returns from their investments. Managing the linkages between public and private sector R & D has become a major management challenge.

The OECD (1997) argues that, as there is an increasing number of institutions with specialized knowledge of very different kinds, the ability to access different sources of knowledge and to apply these to their own needs becomes crucial for the innovativeness of firms. It is the configuration of these institutions and the (resulting) flows of knowledge, according to the OECD, that characterize different national innovation systems and underlie the innovative performance of countries.

The NIS approach suggests that technological innovation is both more frequent and better managed, leading to more substantial national competitive advantage, when the elements of the broader environment surrounding firms' activities are well articulated into a system, than in situations where each element works largely in isolation. Thus, the NIS approach brings a major new form of analysis to industry. According to the OECD, the overall innovation performance of an economy depends not so much on how specific formal institutions (firms, research institutes, universities) perform, but on how they interact with each other as elements of a collective system of knowledge creation and use, and on their interplay with social institutions (such as values, norms, and legal frameworks) (OECD 1997).

A properly functioning NIS underpins the innovative capacities of firms, because the institutional structures provide collectively what firms cannot produce individually. These factors include 'public good' research and the development of a broad range of expertise that is necessary for firms' innovation strategies but cannot be afforded by individual companies. This is especially true in small countries and for small firms that have too few resources to meet the cost of basic research.

Looking at different approaches to finance, management, and education, Patel and Pavitt (1991*a*) distinguish 'dynamic' from 'myopic' national systems of innovation. So, for example, the ability of the education system to produce suitable quantities of skilled technicians, the capacity of the legal system to allow firms to appropriate returns to their innovations, and the nature of the financial system, particularly its ability to fund longer-term, more risky R & D investments, are all important elements of NIS. Each of these issues profoundly affects MTI.

Dynamic national innovative systems are continually changing. The importance of these changes can be seen if we return to our Taiwanese machine tool company in the previous chapter. The company's growth was encouraged by a highly supportive system of research institutions and preferential government financing. But when the company moved to a new stage of technological innovation and arrived at the technological forefront, it developed new requirements from the NIS. It now needed access to basic science and funding for long-term, more speculative R & D. In order to assist this type of firm, the challenge for the Taiwanese NIS is either to change the strategies of existing, successful, institutions, or to try to create new ones.

Regional systems of innovation

The geographical proximity of firms developing and using similar and related products and technologies produces positive sum gains for business and innovation. Since the 1920s, when economist Alfred Marshall showed the importance of 'industrial districts' in providing various supports and synergies for firms, a great deal of research has examined the innovation-promoting potential of firms working and competing together in close geographical proximity. Michael Porter (1990), for example, argues that it is the geographical clustering of industries into systems connected by horizontal and vertical relationships, in combination with factor and demand conditions and firm strategies, that creates innovation and international competitiveness. Industrial districts provide similar firms in proximity with 'aggregation benefits' that can be compared to the reduced transaction costs of integrated firms.

Cooke and Morgan (1994: 26) argue that there are two ways that technological innovation is stimulated regionally. First, 'information, knowledge and best practice are rapidly diffused throughout the local milieu, raising the creative capacity of both firms and institutions'. Secondly, uncertainty is reduced through having a better understanding of the results of decisions. Our Indian software firm, for example, benefits significantly from being in the Bangalore region. Not only is it advantageous for its labour market, but it also provides the possibility for the firm actively to manage local software companies in a network where it acts as the systems integrator.

There are broad differences in regional business and innovation systems. Saxenian (1994), for example, compared what she calls 'local industrial systems'

Box 2.3. ***Key characteristics of regional differences in business and innovation systems***

Characteristics	Silicon Valley: network based	Route 128: independent firm based
Type of firm	Specialist producers	Small number of relatively large integrated companies
Labour market	Open	Emphasis on corporate loyalty
Internal communications patterns	Informal, cross functional	Centralized authority, vertical information flows
External communications patterns	Dense social networks, horizontal communications with customers and suppliers	Emphasis on secrecy
Organizational structures	Porous boundaries with trade associations and universities; extensive collaboration	Distinct boundaries
Approach to change	High levels of entrepreneurship and experimentation, emphasis on rapid change	Emphasis on stability
Approach to learning	Collective learning	Self-reliance

Source: Saxenian (1994)

in California's Silicon Valley and Boston's Route 128 and found several key areas of difference (Box. 2.3). Although both of these regions are renowned for their innovativeness, there are marked variations in the business structures and relationships that create it.

Finally, there may be regional differences in the type and level of government support, sources of science, technology and finance, industrial structures, and indeed innovation-supporting cultures. This is seen particularly clearly in China, where the fifty-two new high-technology zones have been highly successful in building high-tech industries precisely because they are exempted from many government restrictions and procedures applied in other regions of the country. The significant role of government in regulating and facilitating firms' activities in globalized markets adds further complexity to the regional analysis of systems of innovation.

Technology systems

Other perspectives on systems of innovation such as the 'technology systems' approach have developed in parallel with the geographical perspectives we have just discussed. These also focus on systems, but emphasize the specific technologies of the constituent parts of national and international industrial structures.

Technological innovation is rarely a discrete, atomistic event. It invariably builds on extant technology or contributes an element to a broader technological system. Successful innovators integrate their operations with the technological systems of which they are a part. The 'technology systems' approach is in some ways similar to that of national and regional innovation systems but differs in a number of ways. First, the systems are defined by technology rather than national boundaries and, although they are affected by national or regional culture and institutions, they can also be international in nature. Secondly, technological systems vary in character and extent within nations. Thus there are different technological capabilities within nations: Japan is strong in electronics and relatively weak in pharmaceuticals, whilst Britain is the reverse. Thirdly, this systems approach emphasizes technology diffusion and use rather than creation. It is particularly valuable when considering the existence of public-sector research and 'bridging institutions' for the development of different technologies (Carlsson 1994).

The technological systems approach of Carlsson and Stankiewicz (1991) combines both geographical and technological elements and refers to groups of firms within a geographical boundary. They define a technological system as a 'network of agents interacting in a specific economic industrial area [working within] a particular institutional infrastructure'.

The complex product and systems (CoPS) approach, mentioned in Chapter 1 (Hobday 1998), provides another form of analysis. CoPS are essentially large-scale, highly integrated projects and their very complexity points to a number of interesting challenges in, for example, mastering systems achitecture, managing knowledge accumulation in one-off projects, building trust in temporary coalitions of firms, and learning when project teams are often disbanded and quickly realigned. These complex products, which are now designed, produced, and operated as discrete or small batch projects, have led to a growth in firms specializing in project-based activities (Gann and Salter 1998).

Considerations of business and innovation systems add considerably to the complexity of managing MTI. The challenges for MTI are to understand fully the systems of which firms are a part, and to integrate firms' activities into these systems effectively. To relate this discussion to our earlier example of the British pump firm, the firm faces diverse options for its future, which can be clarified with analysis of systems. It can concentrate on simply being a manufacturer of a single product, the pump. Or it can integrate its pumps into larger systems, comprising valves, controllers, gauges, and sensors, which might constitute a subsystem of a larger system, such as a building, a ship, or a refinery. At the most ambitious it could become a systems-integrator itself, taking responsibility for the design and manufacture of the whole system. This decision should be based on a clear understanding of its position within its business system (its relationships with suppliers, customers, and sources of research and finance) and its technology systems (how is the system configured and where does the greatest value lie in being a part of the system).

Networks, complexes, clusters, and chains

Networks are a form of business and innovation system. One of the key features of innovation is the extent to which it involves complicated information and technology flows. The possibility of successful innovation is enhanced through the regular use of multiple channels of communication. Networks of one sort or another are powerful mechanisms for communication. Definitions of network vary, but here they are considered to be an open system of interconnected firms and institutions with related interests (see Castells 1996). Networks offer a rich web of channels, many of them informal, and have the advantage of high source credibility—experiences and ideas arising from within the network are much more likely to be believed and acted upon than those emerging from outside (Dodgson and Bessant 1996).

The strengths of networks (especially for smaller firms) are that they offer a way of bridging gaps between what firms do and what is possible (best practice). Networks can enable the sharing of resources—for example, specialist equipment or R & D projects where the costs and risks of investment to any individual firms would be prohibitive. In addition, networks create the possibility for extensive self-help through experience sharing and learning. Cooperative networks in Europe have enabled small-scale industry to compete successfully in global markets through involvement in collaboration (Best 1990). The Italian furniture industry, for instance, is the world export leader, yet the average firm size is below ten employees.

Freeman (1991: 512) argues that networking should not primarily be explained by reference to costs, but rather in terms of strategic behaviour, appropriation of knowledge, technological complementarity, and sociological factors such as trust, ethics, and confidence in the cooperativeness of others. In addition to the positive benefits, networks can also have negative consequences. The network model of innovation may limit participating firms' access to 'complementary assets' (Hobday 1994) and hence their ability to achieve full commercial returns on innovative activity. They might also have some features of cartels and conceivably exclude possible new entrants, with negative consequences for competition.

A useful way of thinking about innovation systems which incorporates the public sector is Marceau's conception of *complexes*. The analysis of complexes emerges from the work of a group of academics from small countries such as the Netherlands, Denmark, and Australia. Their work derives from a concern to integrate the strong welfare element in these economic systems, the recognition of the considerable importance of public- (rather than private-) sector R & D in these countries, with few major industrial companies and home-based multinationals, and the low absolute level of funding for R & D (Marceau 1994).

The complex is analysed as a network of cooperation between producers (industrial firms), public-sector research organizations, users (usually other firms), and regulators (at different levels of government). For example, innovation in the construction industry complex is affected by the regulations concerning environmental impact and health and safety, and, apart from the influences of companies (materials suppliers, architects, builders), agencies such as planning authorities are also deeply involved (see Gann, 1994). In the healthcare-complex, government purchasing decisions, hospital funding arrangements, and the statutory obligations of health authorities all profoundly affect the level of innovation in pharmaceuticals and medical equipment.

Marceau points to four values of the 'complex' lens in considering innovation systems.

- It contains a specific role for government authorities, which may act directly to assist the complex rather than being reduced to providing general infrastructure or targeting particular companies.
- It indicates the central importance of public R & D facilities (the source of many new technology-based companies, such as the US biotechnology company described in Chapter 1).
- The approach allows the analyst to pinpoint weaknesses in the complex much more clearly and to devise policies for plugging gaps. These weaknesses may lie not in the 'obvious' industrial participants, but in public institutions and policies not usually considered in the overall framework of innovation support.
- It can identify lead organizations—such as hospitals—which, once encouraged towards innovation, can have the knock-on effect of further developing the complex.

Clusters have been defined in a variety of ways. Some focus on the 'horizontal' nature of relationships between small and medium-sized firms that both compete and collaborate with each other. Others would see the relationships between large firms and their core suppliers as leading to clustering in many cases. These are essentially hierarchical relationships which happen to involve inter-firm rather than intra-firm relationships. The equality of relationships between firms found in such clusters derives from the technological interdependence of a group of large and often international firms.

Some observers, most notably Porter (1990), have allocated great importance to the presence of demanding customers as stimulants to innovation in different clusters. Indeed, more generally there is considerable evidence that, in some industries, clients are the critical elements in the development of new products. The biomedical industry is one well known example, where surgeons play a key product design and development role (see von Hippel 1988).

There is some debate as to whether the firms involved in clusters are in the same or related industries. In the wool textile cluster in Prato, Italy, for example, there are both textile companies and the engineering firms that make textile equipment. Similarly, in the Finnish forest cluster, machinery manufacturers are an essential aspect of the cluster's success. The cluster includes both paper manufacturers and the emerging firms that clean up after the paper processes. In that cluster the 'forest' is the key link between the economic activities. In other work, clusters are more strictly defined as parts of an industry (all making leather goods or ceramic tiles), but linked through their inputs to different activities in the production chain. Our Indian software company clearly belongs in a software cluster. Its decision about location and current and future business strategy depends entirely upon its relationship with local users and suppliers of software.

Chains involve links between suppliers and users. One of the primary ways companies learn is by interacting with their suppliers and clients in what have between called 'user–producer relations'. The intensity of these relations varies considerably between industries but seems to be especially important in knowledge-intensive and fast-growing sectors. Chain relations are also important in traditional manufacturing areas. For example, systemic links between users and suppliers in the automotive industry are known to be particularly important for innovation (Womack *et al.* 1990). In Germany the component supplier, Bosch, is the major source of technological change in the car industry (seen most recently in its development of the stabilizing system for all Mercedes cars). A key factor determining the possibilities for innovation in that industry is therefore the closeness and quality of the links between the car assemblers and Bosch.

THE CHANGING NATURE OF MANAGEMENT

New industrial structures and organizational forms require new ways of managing. Innovative, flexible, and imaginative management is needed to deal with the wide range of challenges facing firms and governments in technology, organization, finance, skills, and training, and in their increasingly complex and intimate external links. To deal with some of the challenges outlined above, companies are using many of the practices of what shall be called the *new management paradigm*, aiming to produce more openly communicative and flexible 'learning' organization structures.[1]

[1] This section is based on discussions with, and the work of, a number of management thinkers and practitioners, in particular Stewart Clegg, Gerard Fairtlough, and David Karpin.

There are some deeply pessimistic views about the future of management, typified by Anthony Sampson's analysis of the changing nature of corporate life. Sampson (1995) argues that past certainty of occupation and security of employment are shattered. His 'company man' is insecure, powerless, and totally dominated by overpaid, unaccountable, and centralized top managers, who are themselves helpless in the face of increasingly aggressive corporate raiders. 'Management' is associated with reduced autonomy, increased control, and an inability to defend valuable, hard-earned, and unappreciated assets against dangerous external threats. This view of contemporary corporate management is exaggerated and builds its case on examples of old-style management rather than on emerging modern management practices.

The old, 'scientific management', command and control, models of management are no longer considered to be effective or efficient. What matters to new-paradigm managers is attracting and retaining intelligent, highly motivated, empowered staff and organizing them into coherent and cohesive work groups and networks. Good management encourages high-trust organizations where devolved authority and responsibility ensure fairer and more transparent decisions about resource allocation. Coordination and communication occur as a natural feature of the organization's structure and culture rather than by management fiat. Alternatives to 'bottom-line' evaluations of success are becoming more common, as lessons are learned about the value of models of management more oriented towards growth and stakeholder satisfaction.

Some of the major contrasts between the old and new paradigms of management are shown in Box 2.4. All these issues will be discussed in following chapters, but four of the most germane—strategy, learning, knowledge, and trust—will be raised briefly here.

Strategy

Nowhere are changes in management clearer than in the role of strategic management. As Henry Mintzberg (1994) points out, in the old style of management, strategy meant the development of plans and prescriptions that were to be followed slavishly. In the new style of management, value lies in the process of developing a strategy such that all stakeholders are involved in assessing, understanding, and defining the firm's distinctive competencies. The aim is to define the activities of the firm such that it has a clear identity and purpose in a complex and changing world of seemingly infinite threats and opportunities. These identities and purposes derive in many instances from historical adherence to core values underpinning the organization, and the vision of what the company wants to be as a business (Collins and Porras 1994).

Having the organizational structures and processes that encourage participation in strategic formulation, and ensuring that internal organization facilitates communication across functional and other boundaries, also helps the efficient implementation of strategy. Herein lies a major purpose of strategy.

Box 2.4. ***Two paradigms of management***

Old Paradigm	New Paradigm
Organizational discipline	Organizational learning
Rigid organization	Flexible organization
Low trust	High trust
Command and control	Empowerment
Hierarchies	Markets and networks
Strategy as prescriptive plans	Strategy as process, building consensus, and unity
Not-invented-here syndrome	Receptivity to external inputs
Technology driven by strategic business units	Technology driven by core competencies
Functional structures	Business process structures
Knowledge is periodically useful and resides in few staff	Knowledge is a key source of competitiveness and its creation and diffusion is encouraged throughout the firm

The innovations and changes that occur in all successful organizations are disruptive and dysfunctional unless their purpose is well articulated and communicated, they are coordinated, and those affected by them feel a sense of ownership and influence over their nature and outcomes. Strategy in this sense helps build consensus and unity; it empowers rather than directs.

Strategy determines the extent to which the firm is prepared to undertake risky investments. As we shall see in subsequent chapters, there are advantages in having a balanced research portfolio with some high-risk/high-return projects. It is incumbent upon managers with strategic responsibilities to ensure that the longer term is not ignored when short-term pressures are strong. Our pump firm's expensive, and hence, risky investment in CAD showed a long-term perspective and was based on a strategic decision on what the firm wished to be in the future. The risk in entrepreneurial start-up firms, like our biotechnology firm, is high. As we saw, it initially focused on the wrong product. Acceptance of that level of risk is determined by the strategy of the company's owners (in this case, the professional risk-taker, the venture capitalist). Some of the major elements of strategy will be examined in consideration of technology strategy in Chapter 6.

Learning

Learning is such an essential element of corporate competitiveness that it has led a number of observers to feature it as the defining characteristic of successful firms: the 'learning firm' (Senge 1990; Garvin 1993; Howard 1993). Learning firms focus their efforts in order continually to transform themselves with the aim of building and enhancing their competencies. That is, learning firms learn *purposefully* and *productively*. Learning occurs by means of investments in resources, and policies towards employees, other stakeholders, and other firms with which they have business relationships. There is a large and growing literature on learning in organizations and firms.[2] This literature points to the analytical complexity and multifaceted nature of learning, but a number of features are clear.

- Learning is a long-term activity and needs to be directed by cogent strategies.
- It is expensive.
- It can be encouraged by a variety of external linkages. Our Japanese R & D lab, for example, learns extensively from its overseas links.
- The organizational challenge for firms is to transfer individual learning into group practices and corporate routines.

A feature of technological learning lies in the way its development is cumulative and path-dependent. That is, as far as a company is concerned, history matters and what you do today and tomorrow depends in large part upon what you did in the past. Learning is a key element of MTI, and is examined in greater detail in Chapters 6 and 7.

Knowledge

Like learning, knowledge has been described as a central defining characteristic of firms and their ability to compete. According to Kogut and Zander (1993), 'firms are social communities that specialize in the creation and internal transfer of knowledge'. It is argued to be increasingly important, as capital- and labour-intensive firms, and routine work, are replaced by knowledge-intensive firms and activity, and knowledge work (Starbuck 1992). There is a large number of analyses of the nature of knowledge, with Maschlup (1982) providing the most comprehensive. A detailed analysis of the various components of knowledge management within an organization is provided by Johnston and Blumentritt (1998) (see Box 2.5). Four aspects of knowledge that affect MTI will be outlined here.

[2] Dodgson (1993*c*) has a review of some of this literature.

Box 2.5. ***A typology of knowledge management***

Knowledge *identification*: the process of locating and recognizing knowledge that is relevant to the organization.

- This might be the result of a specific or a routine search, or contact with a member of a network, or arise through serendipity.
- Regardless, the process depends on a selection mechanism that can identify relevance.

Knowledge *acquisition*: the process of obtaining knowledge previously not available to the organization in a form available for exploitation.

- This procedure could vary from simply obtaining public knowledge, as from the literature, to conducting a survey of customers.
- A great deal of knowledge acquisition occurs at the individual level.

Knowledge *generation*: the process of creating new knowledge within an organization, whether through traditional R & D, or through the linking of previously separate information (e.g. on customer needs and technology capabilities).

- Much of this occurs at the individual or team level.

Knowledge *validation*: the process of determining both the accuracy and the value of knowledge, from the perspective of the organization.

Knowledge *capture*: the process by which the organization gains control over particular knowledge.

- This process may involve the purchase of rights to certain proprietary knowledge from another firm, or it may be the transformation of the personal knowledge of a member of staff, or the output of a team, into an explicit organizational resource.
- Only with capture can the knowledge become an exploitable asset of the organization.

Knowledge *diffusion*: the process of spreading knowledge through an organization, and to targets outside the organization (e.g. customers or regulators).

- This process may take place in a relatively informal manner, through conversation and discussions, or through more explicit processes designed to ensure that relevant staff and sections are informed about particular knowledge assets considered valuable for a particular objective.

Knowledge *embodiment*: the process of transforming knowledge within the organization into a form in which its value becomes evident inside and outside the organization.

- This process may involve training, establishment of new procedures, or absorption into the organizational culture.
- Diffusion can occur without embodiment, usually resulting in the knowledge asset being less readily available or used.

Knowledge *realization*: the process of identifying, or becoming aware of, knowledge assets held within an organization (as such a subset of knowledge identification) and managing them to achieve the maximum value-added to the company and customers.

- The development and mining of technology or knowledge platforms is one approach to knowledge realization.

Knowledge *utilization/application*: the process of deliberately and intentionally using knowledge to pursue a specific objective.

Source: Johnston and Blumentritt (1998).

First, knowledge is something that needs to be managed. This includes all the aspects of knowledge described by Johnston and Blumentritt (1998). According to Leonard-Barton (1995), knowledge-building activities are crucial elements in the definition of core technological capabilities, which are in turn integral elements of competitive advantages. These activities include shared problem solving, experimentation, prototyping, importing and absorbing technological and market information, and implementing and integrating new technical processes and tools. As knowledge is an asset, it is something that needs to be accounted for, and a number of efforts are being made to develop procedures for measuring it (Sveiby 1997).

Secondly, knowledge has some distinctive characteristics when it is considered as something that is marketable. Economists describe knowledge as being *non-rivalous* (that is, once it is produced it can be reused by others) and *non-excludable* (that is, it is difficult to protect once in the public domain). It is also *indivisible*—that is, it must be aggregated on a certain minimum scale to form a coherent picture before it is applied (Johnston 1998). These features of knowledge have major implications for the management of intellectual property, which will be discussed in Chapter 8.[3]

Thirdly, knowledge is not just information that can be stored, transported, and accessed electronically. It includes subjective, considered, and personal assessments of the value, meaning, and use of information (Nevis *et al.* 1997). Knowledge is, therefore, personally and socially embedded, and this is an issue challenging managers relying on electronic media as sources of knowledge advancement.

Fourthly, it is necessary to distinguish between codified and non-codified (or tacit) knowledge. This distinction, identified by Polanyi (1967), separates knowledge that is easily communicated by, for example, being written down in papers or blueprints, and knowledge that cannot readily be described, including craft knowledge, and that is learnable only by observation and imitation.

[3] Some sociologists of science question this view of knowledge and argue that knowledge is much more 'sticky', making its transfer harder.

The differences between tacit and explicit knowledge, and the links between them, provide the basis for Nonaka and Takeuchi's (1995) approach. Tacit knowledge, they argue, is developed in the individual, and they identify a 'knowledge spiral' by which it is translated into explicit forms of knowledge. Nonaka and Takeuchi develop a view of a new kind of organization—the hypertext organization—which develops new knowledge by the conversion of tacit into codified knowledge.

Trust

Trust facilitates learning within the firm and in external relationships between organizations. Fukuyama (1995) suggests that national prosperity depends more on communitarian behaviour of shared values and sociability than on the rational selfish interest ascribed to humankind by so many economists. This emphasizes, of course, the importance of the relational approach to national innovation systems. Trust has been seen as a key feature in facilitating continuing relationships between firms (Macauley 1963; Arrow 1975), and it can underpin economic relationships, particularly in the form of buyer–supplier relations (Sako 1992; Sako and Helper 1998), networks (Saxenian 1991; Sabel 1993) and technological collaboration (Dodgson 1993*b*).

An example of the extent and benefit of trust is seen in Saxenian's (1991) study of successful Silicon Valley firms. She illustrates how these firms exchange sensitive information concerning business plans, sales forecasts, and costs, and have a mutual commitment to long-term relationships. Saxenian (1991: 428) argues that this involves 'relationships with suppliers as involving personal and moral commitments which transcend the expectation of simple business relationships'.

In addition to trust being the cement in the relationship between firms, it is also essential within firms (Fox 1974). Trust is necessary, according to Sabel (1993: 332), because,

> as markets become more volatile and fragmented, technological change more rapid, and product life cycles correspondingly shorter, it is too costly and time consuming to perfect the design of new products and translate those designs into simply executed steps. Those formerly charged with the execution of plans—technicians, blue-collar workers, outside suppliers—must now elaborate indicative instructions, transforming the final design in the very act of executing it'.

A brief note on the nature and extent of trust in industry

Trust, like learning, is by no means a straightforward concept. Sako (1992: 377) argues trust to be 'a state of mind, an expectation held by one trading partner about another, that the other will behave in a predictable and mutually acceptable manner'. She argues there are three types of trust to be distinguished. 'Contractual trust' exists such that each partner adheres to agreements, and keeps promises. 'Competence trust' concerns the expectation that a trading partner will perform his role competently. 'Goodwill trust' refers to mutual expectations of open commitment to each other. 'Someone who is worthy of "goodwill trust" is dependable and can be credited with high discretion, as he can be expected to take initiative while refraining from unfair advantage taking . . . trading partners are committed to take initiatives (or exercise discretion) to exploit new opportunities over and above what was explicitly promised' (Sako 1992: 379). Sako's studies reveal the advantages firms (primarily Japanese firms) enjoy in subcontracting relationships that are not at arm's length, but rather are 'obligational'. The form of the subcontracting relationships adopted by our Indian software company will be critical to the success of its network.

Trust is, therefore, central to a number of elements of the innovation process, in particular inter-firm relationships and creative, committed organizations. But how trusting are inter- and intra-firm relationships? Hofstede's (1980) analysis of comparative national cultures examined, amongst other issues, the question of individualism versus collectivism. Individualistic cultures are loosely knit and people look after themselves and their immediate families only. Collectivist cultures are characterized by a tight social framework where groups enjoy strong loyalties and are expected to demonstrate substantial mutual commitments. Hofstede's (1980) analysis of thirty-nine countries found that the most individualistic was the United States. The Asian economies, by contrast, had generally low individualism. Whilst not equating collectivism with trust, this indicator might point to a propensity of many Western countries to avoid cohesive networks and groups, while this is one of the features that has provided many Asian economies (such as Japan) with a considerable source of strength.

High-trust networks exist in many countries and industries. In Australia (also individualistic according to Hofstede) there is a very cohesive network in the mining industry, and it is one of the most successful mining industries in the world (se Fig. 2.2). Such networks cannot exist without high levels of trust and the award of considerable discretion to other parties. Contracts encompassing the complex and diverse activities undertaken within the network would be virtually impossible to draft, let alone police. The efficiency of the system might also be enhanced by the existence of goodwill trust where parties offer more than was expected of them in the expectation that the favour will be reciprocated in the future. As Quinn (1992: 386–7) argues, 'in successful ventures, trust was consciously nurtured through three rules: reliability, cooperativeness, and openness . . . Without trust engendered by these mechanisms, the smallest misunderstandings, which always occur, can quickly become large issues rather than a pattern of small successes that help re-inforce trust itself'.

Throughout his studies of outsourcing in the Australian mining industry, Quinn was continually surprised by how little was actually written down in

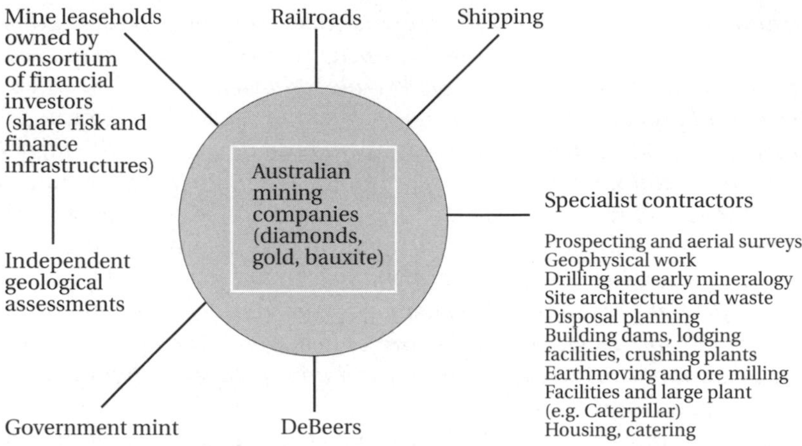

Fig. 2.2. Networking relationships in Australian mining
Source: Quinn (1992).

formal contracts (outside prices and broad parameters of performance). Such networks can also possess self-regulating aspects inasmuch as sanctions can be applied if a company misbehaves—that is, it can be excluded from further work with the other firms in the network.

The challenges for MTI in all the case-study companies described in Chapter 1 are to move away from the pessimistic Sampsonite view of management, and towards the new paradigm of management. Whether working in a manufacturing firm in the United Kingdom or Taiwan, a science-based firm in the United States, a Japanese multinational, or an Indian services company, the challenge is to use strategies that focus on learning and knowledge and are based on trust. It is the ability of firms to be flexible, to empower their workers so as to encourage the creation and diffusion of knowledge, and to use strategies to develop unity of purpose, that will make these firms effective organizations and provide their future competitiveness.

THE CHANGING NATURE OF THE INNOVATION PROCESS

Technological innovation is such a complex process that over twenty-five years or so analysts have developed several approaches to explaining its nature and how it works. Researchers have posed a series of questions for analysing the type of innovation activity. They ask, for example, whether the innovation

- is *radical* or *incremental* (Freeman 1974);
- is *continuous* or *discontinuous*—that is, whether it affects existing ways of doing things (Tushman and Anderson 1986);
- has *transilience* in that it affects existing ways of doing things (Abernathy and Clark 1985);
- changes over *life cycles* (Abernathy and Utterback 1978);
- is *modular*—that is, occurs in components and subsystems without addressing the system of which they are a part—or *architectural*—that is, attempts systemic improvements without great attention to its component parts (Henderson and Clark 1990);
- results in the emergence of *dominant designs* (Abernathy and Utterback 1978);
- is *sustaining* or *disruptive* (Christensen 1997; see also Chapter 7).

Other approaches consider the sources and nature of innovation more broadly, and these can be categorized into five generations of thinking (Rothwell 1992; see Box 2.6 below).

The first, prevalent during the 1950s and 1960s, was the *science-push* approach. This approach assumed that innovation was a linear process, beginning with scientific discovery, passing through invention, engineering, and manufacturing activities, and ending with the marketing of a new product or process. Until the 1970s, many government policy-makers and managers of major industrial companies accepted the view that a new product or process is the result of discoveries in basic science, brought to the attention of the parent organization by its research staff for possible commercial applications. In this model there are no forms of feedback. The model was rapidly shown to apply only to relatively simple forms of product, such as petrochemicals.

From the early to mid-1960s a second linear model of innovation was adopted by public policy-makers and industrial managers in advanced capitalist economies. This was the *demand-pull* model. In this model, innovations derive from a perceived demand which then influences the direction and rate of technology development. Kamien and Schwartz (1975: 35) argue that in this model innovations are induced by the departments that deal directly with customers, who indicate problems with a design or suggest possible new areas for investigation. The solutions to any problems raised are provided by research staff.

Many commentators now see both linear models of innovations as oversimplified (Rothwell 1992; Steinmueller 1994). Rothwell (1992), for example, uses the case of the biotechnology industry to show that not only does the development of few products conform to science-push models but also that at an industry-wide level the importance of science-push and demand-pull may vary during different phases in the innovation process.

The third model, the *coupling* model, integrating both supply-push and demand-pull, was centred around an interaction process where innovation was regarded as a 'logically sequential, though not necessarily continuous

Box 2.6. ***The five generations of innovation process***

FIRST GENERATION

Technology push. Simple linear sequential process. Emphasis on R & D. The market is a receptacle for the fruits of R & D.

SECOND GENERATION

Need pull. Simple linear sequential process. Emphasis on marketing. The market is the source of ideas for directing R & D. R & D has a reactive role.

THIRD GENERATION

Coupling model. Sequential, but with feedback loops. Push or pull or push–pull combinations. R & D and marketing more in balance. Emphasis on integration at the R & D/marketing interface.

FOURTH GENERATION

Integrated model. Parallel development with integrated development teams. Strong upstream supplier linkages. Close coupling with leading-edge customers. Emphasis on integration between R & D and manufacturing (design for makeability). Horizontal collaboration (joint ventures, etc.).

FIFTH GENERATION

Systems integration and networking model (*SIN*). Fully integrated parallel development. Use of expert systems and simulation modelling in R & D. Strong linkages with leading-edge customers ('customer focus' at the forefront of strategy). Strategic integration with primary suppliers including co-development of new products and linked CAD systems. Horizontal linkages, including joint ventures, collaborative research groupings, collaborative marketing arrangements, etc. Emphasis on corporate flexibility and speed of development (time-based strategy). Increased focus on quality and other non-price factors.

Source: Rothwell (1992).

process' (Rothwell and Zegveld 1985: 50). The emphasis in this model is on the feedback effects between the downstream and upstream phases of the earlier linear models. The stages in the process are seen as separate but interactive.

New models (the 'fourth- and fifth-generation' innovation models, as Rothwell calls them) have incorporated the feedback processes operating within and between firms. The high level of integration between various elements of the firm in innovation is captured in the fourth-generation, '*chain-linked* model' of Kline and Rosenberg (1986), which shows the complex iterations, feedback loops, and interrelationships between marketing, R & D, manufacturing, and distribution in the innovation process.

The fifth-generation innovation process includes the growing strategic and technological integration between different organizations inside and outside the firm, the way these are being enhanced by the 'automation' of the innovation process, and the use of new organizational techniques, such as parallel rather than sequential development. It moves away from the 'silos' of functional structures towards organization according with business processes. Box 2.6 shows how conceptualization of the innovation process has changed over various generations.

The fifth-generation innovation process

While the analytical origins of the fifth-generation innovation process differs from the macro-perspective of the fifth wave of technological development, the challenges facing management in both are similar, particularly in respect to dealing with high levels of risk and uncertainty. The major aspects of the fifth-generation innovation process are outlined in Fig. 2.3. Within the firm we see increasing concern to find the organizational forms and practices and skill balances that enable the maximum flexibility and responsiveness to deal with unpredictable and turbulent markets. Some of these issues have already been discussed in relation to the new management paradigm, and others, like lean thinking, will be discussed in Chapter 5. The value-creating activities of the firm are linked with suppliers and customers, and all the technological activities in the firm are directed by increasingly coherent and effective technology strategies (see Chapter 6). Two important features of the fifth-generation innovation process are the increasing extent of strategic and technology integration.

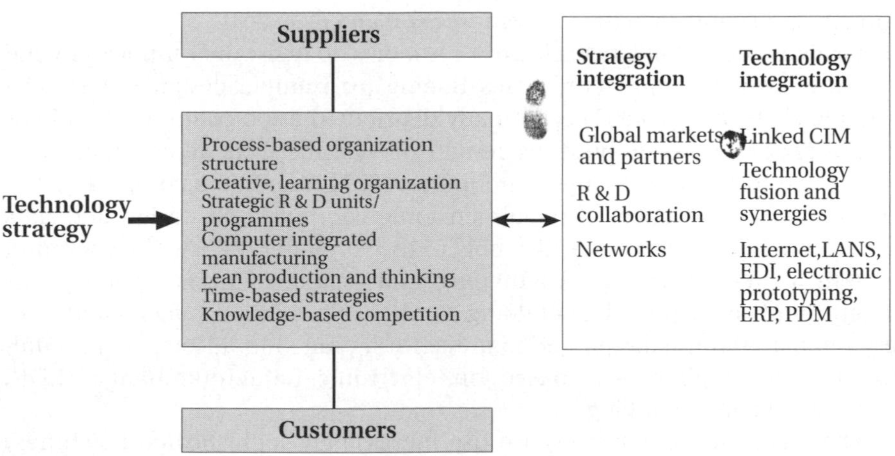

Fig. 2.3. Towards the fifth-generation innovation process

Strategic integration between firms is increasingly global and occurs across technological, market, and financial areas. When Boeing designed its 777 jet, it closely involved its customers and suppliers. Boeing created what came to be known as the Gang of Eight, comprising eight international airline customers who met over twelve months to help specify the needs for the new aircraft. Reflecting its decision to purchase thirty-four of the new aeroplanes before they had even been designed, was the fact that one of these clients, United Airlines, was intimately involved in its configuration (Sabbagh 1996). Boeing also closely involved its suppliers. Important components, such as major parts of the fuselage and the rudder, were subcontracted to Australian and Japanese firms. Engine manufacturers, such as Pratt and Whitney, designed their engines in close conjunction with Boeing.

Since 1995, Ford has operated a Technology Review Center in Dearborn, Michigan. This is a forum for suppliers to demonstrate their technological expertise to Ford engineers. These 'supplier showcases' typically last two days and have been replicated in the United Kingdom and Germany. A major problem for Ford has been integrating the technology of its suppliers and these centres are designed to overcome these problems by giving early feedback on suppliers' technology developments.

Technological integration occurs in various forms. An example would be the hybrid car, running on both electricity and petrol, and involving the merger of electrical and mechanical technologies. Kodama (1995) discusses the increasing prevalence of 'technological fusion'. Thus 'mechatronics' involved the fusion of mechanical technology with electrical and material technologies, and 'optoelectronics' involves the fusion of glass technology with cable and electronic device technologies. He argues that fusion is more than a combination of different technologies but the creation of a new technology where the whole is greater than the sum of the parts. Each fusion 'creates new markets and new growth opportunities in the innovation' (Kodama 1995: 203).

Firms are developing new electronic 'toolkits' to assist their innovation and new product development processes. Boeing, for example, designed the 777 jet completely by computer. The company distributed 2,200 computer terminals to the 777 design team, both inside and outside the firm, all of whom were connected to the world's largest grouping of IBM mainframe computers. The designers used advanced three-dimensional computerized design and data sets, known as electronic pre-assembly in the computer (EPIC). The computer system allowed electronic prototyping and testing of parts, avoiding the lengthy and costly processes of doing so physically. EPIC also facilitated integration between the design and manufacturing function. Electronic procurement with suppliers was based on electronic data interchange (EDI), eliminating paper handling.

Pharmaceutical companies also use increasingly sophisticated machinery in fields such as gene sequencing and combinatorial chemistry. Pfizer has developed automated techniques for rapid screening and dispensing of thousands of chemicals using robots. GM uses what it calls 'math-based design and engineering', which provides digital representations of vehicles for design, engineering, and testing. This system is used routinely in areas such as

- 3-D simulations for modelling vehicle structures, crashworthiness, safety restraint and manufacturing tooling;
- computational fluid dynamics codes for designing and analysing engine combustion systems, transmissions, interior climate control systems, and vehicle aerodynamics;
- high-level systems for automated design of integrated chips and electro-mechanical components;
- virtual reality prototypes for vehicle exteriors, interiors, components and production tooling.

Use of these systems has led to developments in stability control systems, electronically enhanced steering, continuously variable suspensions and engine controls, with virtual-reality prototyping reducing the need for costly physical models (Baker 1997). The electronic toolkit is also very important in the design of large, complex systems such as utilities and communications systems, where it is not usually feasible to test full-scale prototypes.

Gann and Salter (1998: 435) argue that

> simulation and modelling (in complex systems) is, therefore, of great importance in front-end decision-making, planning and execution. Product definition, development, simulation, testing and production usually involve the transfer of knowledge within complex networks of suppliers, as well as a large number of interactions between many different specialists. This includes the need to deal with technical decisions, in which the interdependency between components and subsystems creates the need for an exchange of technical know-how across a range of professional and engineering disciplines.

This increased strategic and technological integration often has as its aim improvement in competitiveness through the timeliness of delivery of goods and services. Time-based strategies of rapid speed-to-market are growing in importance. When Sony developed the camcorder, it believed that it only had a six-month lead on its competitors. One of Toshiba's laptop computer manufacturing plants introduces a new model to the manufacturing line every two weeks. At one stage Nokia introduced a new mobile phone every month for two years. Speed is particularly important in the pharmaceutical industry, where first-to-patent is of crucial competitive importance, and the speed of the development process can provide distinct advantages. It is also important in online services, where being first-to-market assists the development of brand recognition. Fig. 2.4 shows the short product lifetime of various personal stereos (Sanderson and Uzumeri 1995).

To reduce the time it takes to develop new products, digital product data must be presented in a way that can be used effectively by all the different departments of the firm. Software companies, such as SAP, offer enterprise resource planning (ERP) systems that help integrate financial data with design,

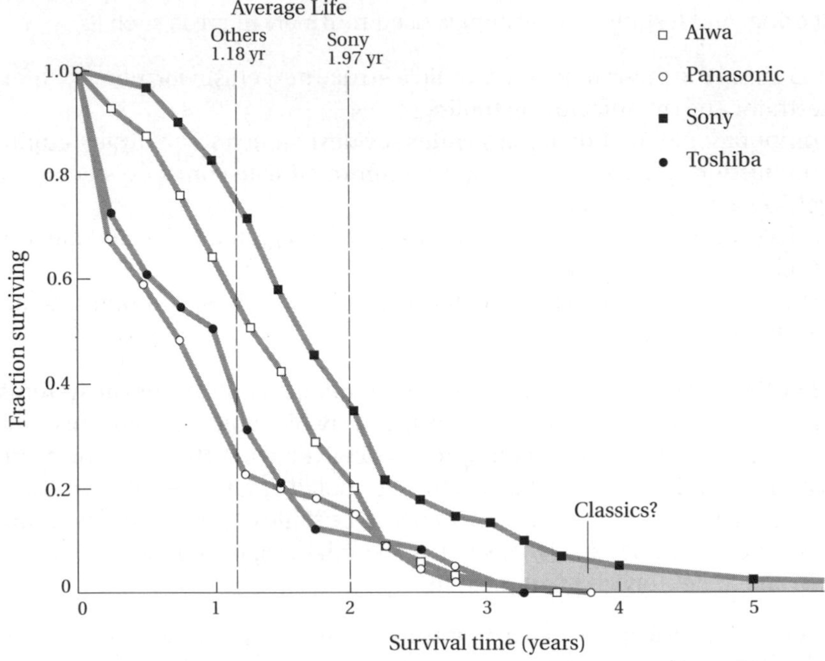

Fig. 2.4. Short product life of personal stereos
Source: Sanderson and Uzumeri (1995).

manufacturing, and inventory data. Although very complex in practice, when it is fully operational ERP enables each department in a firm to have access to the information it requires in a format it understands.

Product data management (PDM) systems can be part of ERP systems. PDM stores all the information and data about products in an easily available format, allowing continual changes to be made in a controlled manner. In his research on Daewoo Motors, Sigurdson (1999) describes how Daewoo expects PDM to enable advanced simulations in the design of new cars, and to become important in the continuing modifications required to deal with national regulations, consumer preferences, and production processes. Daewoo's PDM system already facilitates the development of subsystems, such as complete door systems, and in doing so generates up-front data that can be transferred to later stages for engineering in the production process.

All our case-study companies are affected by aspects of the fifth-generation innovation process. The biotechnology firm has to be particularly closely integrated with its major partner, and its future strategy is closely entwined with it. The Taiwanese machine tool firm and the Japanese R & D laboratory need close affiliations with international research laboratories. The British pump firm requires close strategic integration with other firms involved in the design and manufacture of its systems. The Indian software company aims to act as a systems integrator, coordinating inputs from suppliers to meet customer needs. High levels of technological integration are required of all firms. The biotech firm is using a number of automated systems for gene sequencing and this enables the 'design' of molecules electronically. The pump firm has a CAD/CAM system linked to its suppliers and customers. The machine tool company uses a range of computerized design data bases, enabling it to store all information on existing products and test results and electronic prototyping of new parts. It is planning to integrate its computer systems. The Japanese laboratory uses a wide range of Internet-based communications to link with its internal customers, and external scientific partners. The Indian software company uses computerized project management systems and software writing tools.

A brief note on measuring the economic contribution of technological innovation

Measuring the relationship between technological innovation and economic growth is difficult because of data shortcomings and the complexity of innovation itself (it is not a 'discrete' activity, but rather involves a number of interactions over a lengthy period). As Christopher Freeman (1994) argues, however, while it is difficult to measure the precise contribution of technical change to the growth of industries and countries, no one doubts that innovation is essential to this process.

The OECD is attempting to produce coherent internationally comparative surveys of innovation using the so-called Oslo Manual. *Rather than being able to measure 'innovation'* per se, *most large-scale empirical studies rely on the use of proxy measures of innovation using national data. These are commonly R & D expenditures and numbers of patents, and there are problems with both. R & D might be considered an* input *measure of innovation, but there are problems with what is included as R & D. For example, some firms might consider software development to be R & D, others may not. Small firms often under-record the amount of R & D they undertake, as it might be incorporated within the broader activities of entrepreneurs and key personnel. As a result, their R & D may be under-represented by national surveys. Patents might be considered an* output *measure, but patents are irrelevant in some areas—for example, in some areas of software. Also, nations have different approaches to what is patentable and how long the protection lasts. Despite these shortcomings, by combining R & D and US patenting data, Patel and Pavitt (1991a) argue, it is possible to provide a plausible and consistent picture of technological activities at the world's technological frontier.*

There are many econometric studies of the returns to R & D investment but, like all efforts to measure an activity that is complex, socially based, and long term, they are bedevilled with methodological problems and inadequacies. Many of these studies measure the 'social' rather than the 'private' rate of return, because often the firm (or government agency) that undertakes the research does not accrue all the benefit. Once a technological breakthrough has occurred, it is replicable in one form or another by others who also accrue benefit (this is called 'spillover' in the economics literature).

Amongst the best studies of private returns are those of Griliches (1986), whose firm-level estimates of the gross rate of return to industrial R & D range between 25 and 40 per cent; Odagiri (1985), whose study in Japan estimated returns of 26 per cent; and Mansfield (1988b), whose industry-level analysis shows returns of 38 per cent in Japan and 27 per cent in the United States. Coe and Helpman (1993), in a study of the twenty-two OECD nations plus Israel, show a rate of return of between 100 and 120 per cent. While there is some inconsistency in the econometric research, there is a general consensus that R & D investment produces advantageous rates of return with benefits considerably outweighing costs.

A further form of analysis is that of Cohen and Levinthal (1990), who argue that one of the major outcomes of R & D is the increasing 'absorptive capacity' of the firm—that is, the improved ability of the firm to absorb external knowledge. This product of R & D will be examined in greater detail in Chapter 3.

Despite the measurement difficulties, we do know (as we saw in Chapter 1) that R & D expenditure and patenting activity is positively associated with growth in productivity and exports and that the use of advanced manufacturing technology is linked to increased employment, higher wages, and more secure jobs. Furthermore, the evidence points to a rapidly growing world trade in R & D-intensive products and knowledge-intensive services and to these products and services accounting for an increased proportion of world trade. Those economies that can expand their level of economic activity and exports in high-technology goods and in knowledge-intensive services will be best placed in the global economy.

GLOBALIZATION

Globalization is a broad term encompassing a wide range of issues and developments. It includes changes in corporate strategies in relation to production, marketing, finance, and R & D and, from the corporate perspective, it can be conceived as the search for competitive advantage across national borders. GM, which operates 365 facilities in over fifty countries, defines being global as 'taking advantage of your global resources to do a better job of bringing customer-valued products to market, faster and for less money than your competitors can' (GM Vice-President, Kenneth Baker, 1997). An indicator of the increased level of globalization is seen in the increase in companies' direct foreign investment, increasing from $700 trillion in 1985 to $2,730 trillion in 1996.

There are various drivers behind globalization.

- *Greater participation in, and integration of, world trade.* An increasing number of nations are becoming involved in world trade and are subjecting themselves to the conditions and disciplines required actively to participate in it. The membership of the World Trade Organization (WTO), and its predecessor, the General Agreement on Tariffs and Trade (GATT), increased from sixty-two countries in 1967 to 130 in 1997, with another twenty-nine applying for membership. As part of this expansion, the Asia Pacific has become a centre of international production and exports, and its increasingly affluent markets are important components of world trade. Similarly, but to a lesser extent, the Latin American countries and transforming East European nations have also expanded their international trade links.
- *Liberal government policies.* Governments internationally have followed the paths of deregulation and tariff reductions, encouraged by the WTO, and have lifted the constraints of protectionism and facilitated greater investments from overseas in important areas such as telecommunications and banking.
- *Changing corporate strategies.* Companies are increasingly sophisticated in their international management. Multinational companies had sales of $7 trillion through their foreign affiliates, an amount greater than the world's total exports in 1995 (*The Economist*, 22 Nov. 1997). Multinationals are more prepared, not only to invest overseas, but to form strong partnerships and joint ventures with overseas companies. The Asian strategy of the German chemical company BASF, for example, includes nine joint ventures in China (one shared with Du Pont) and a joint venture in Korea. It is the largest shareholder in the Japanese pharmaceutical company Hokuriku. It is constructing a major manufacturing plant in Malaysia, and is forming a joint venture with Malaysia's largest company, Petronas, to create another. The reasons BASF is investing so heavily in China, despite the political and commercial risks involved, are illuminating and are described by the company in the following manner. 'The risk to BASF of missing out on the big opportunities offered by China can be regarded to be much greater than any risk that might arise from unfavourable political developments in that region. A successful investment in China will make substantial contributions to the long-term stability and profitability of the BASF Group as a whole' (BASF 1997). Direct foreign investment can occasionally involve the transfer of technological and managerial expertise, which can then diffuse through the economy and improve the nation's capacity to operate in global markets.
- *Creation of global capital markets.* Liberalized capital markets have encouraged cross-border capital movements. Capital is increasingly available internationally, as major financial markets such as Wall Street and the City of London become more global in orientation. Daily foreign-exchange

turnover has increased from $15 billion in 1973 to $1.2 trillion in 1995. Cross-border sales and purchases of bonds and equities have risen from the equivalent of 9 per cent of US GDP in 1980 to 164 per cent in 1996 (*The Economist*, 18 Oct. 1997).
- *Capacities of information and communications technologies.* Technologies such as satellite and broadband communications systems and the Internet have provided firms with the potential to facilitate communications across borders. The cost of a three-minute telephone call between New York and London has fallen from $300.00 in 1930 (in 1996 dollars) to $1.00 in 1997 (*The Economist*, 18 Oct. 1997). General Motors' common computer and networking protocols ('math-based modelling' utilizing common mathematical systems and representations) is argued to be critical for dealing with the complexity and scope of product development on a global scale (Baker 1997).
- *Increasing market homogeneity.* While it is important not to exaggerate the extent to which cultures and markets are becoming alike, the ubiquity of certain brands and images—Disney, Coca-Cola, Sony, McDonalds, Nike—reveal a certain convergence in international tastes and experiences.

In addition to these factors, the *creation, use, and sale of technology* provide a major reason why firms are globalizing. The United Nation's 1997 *World Investment Report* estimates that 70 per cent of all international royalties on technology involve payments between parent firms and their foreign affiliates, showing that multinationals play a key role in disseminating technology around the globe.

Early research on the international diffusion of technology argued that firms developed technology in their home markets, then exported their products, which led to overseas production and eventually to some simple R & D activities to amend products to local tastes and requirements (Vernon 1966). It was subsequently recognized that some firms, particularly in high-technology sectors, undertook R & D concurrently in different markets (Vernon 1979). In the 1990s, firms became more globalized in their orientation towards technology development and use. There continues to be debate about the extent to which R & D has become internationalized. Some argue that R & D is such an important strategic activity that it essentially remains controlled in corporations' home nations (Patel and Pavitt 1991*b*). Others argue that increasing amounts of R & D derive from foreign investment.

The following research findings indicate the extent to which R & D and patenting are globalized.

- In a study for the UK Treasury it was found that 45 per cent of UK R & D was undertaken overseas (Martin and Salter 1996).
- More than 15 per cent of total US industrial R & D expenditure is undertaken by foreign firms (Florida 1997).
- At the end of 1994, more than 300 foreign companies owned more than 645 R & D centres in the United States (Dalton and Serapio 1995).

- 41 per cent of European chemical and pharmaceutical firms' R & D employees in biotechnology operate outside Europe (Senker *et al.* 1996).
- 30 per cent of Siemens' and 22 per cent of IBM's research employees work outside these companies' home countries. Companies such as Nortel, Nokia, Lucent Technologies, and Dow Chemicals have over ten research centres overseas.
- 45 per cent of US patents are registered from abroad, and 45 per cent of European Patent Office patent applications come from non-member countries (Archibugi and Michie 1995).

A case of just how globalized technology can be is provided by a Daewoo telecommunications product.[4] Daewoo Telecom, the Korean company, developed a new switching system for the telecommunications network in Korea. The development cost roughly $25 million over three years, and involved around 350 engineers. Much of the background work was done in Korea; however, substantial software coding was done in New Delhi. An important technical requirement in which Daewoo had no expertise, the advanced intelligent network, was developed in Daewoo's US Telecom Research Center. This centre is headed by an American, previously from Bell Labs, who sourced some of the required technology from a company located in Princeton, New Jersey. The product is now being considered for purchase by Russia, Uzbekistan, Ukraine, Kazakstan, India, and Myanmar.

In all discussions about the extent of globalization of R & D, it is clear that there are broad differences between countries. European firms tend to be more internationalized in their overseas R & D investments than US firms, which, in turn, are more globalized than Japanese firms (Roberts 1994). Firms based in smaller countries, like the Scandinavian nations and the Netherlands, do not have a large domestic market, and if they are to undertake substantial R & D projects, they have to exploit the results internationally (Archibugi and Michie 1995).

One of the major explanations of the greater globalization of technology is the greater balance in international contributions to R & D. The US share of the seven largest industrial nations' R & D reduced from 70 per cent in 1960 to less than 50 per cent in 1994. Countries such as Korea and Taiwan, which have grown rapidly in their technological expertise, are undertaking greater levels of R & D and patenting in the United States. According to one study, total R & D expenditure in Asia (including China and India) will exceed expenditure in the United States by 2005 (Sheehan *et al.* 1995). A study of multinational investment in R & D in India showed that a significant proportion was involved in higher-order R & D, related to the development of new products and processes (P. Reddy 1997). Another reason for increasing globalization is that governments invariably welcome and encourage overseas R & D investment and may have

[4] The source of this information is Jon Sigurdson, European Institute of Japanese Studies, Stockholm School of Economics.

incentives to encourage that investment. Singapore, a classic example in this regard, offers a range of government grants and incentives. Despite some concerns voiced in the United States about the negative consequences of overseas companies accessing US science cheaply, a number of reports have revealed the benefits of overseas R & D investment, particularly through the encouragement of highly skilled employment (Dalton and Serapio 1995). Whilst this globalization of technology is occurring, it is important to note that it is primarily in Japan, Europe, and the United States where it is most advanced.

Another aspect of globalization is the way science is increasingly internationalized, with a large proportion of academic publications being derived from international collaborations, particularly in basic research (Bourke and Butler 1995).

Firms have different motives for globalizing their technological activities. These are described as: (*a*) global *exploitation* of technology, (*b*) global technological *collaboration*, and (*c*) global *generation* of technology (Archibugi and Michie 1995). The global exploitation of technology is considered to be akin to export flows, and is primarily the incorporation of technology embodied in products and services. This would occur if our Taiwanese machine tool company or British pump firm sold its products in Japan in configurations that met national standards requirements. There may be some limited MTI considerations in respect of adaptation of these products to local requirements either at a local level or in the home market. Global technological collaboration and its management would be a major issue if the US biotechnology firm began collaborating with a European pharmaceutical firm. This will be considered in detail in Chapter 7. Global generation of technology is probably occurring to a much lower extent than the other categories (Casson 1991; Archibugi and Michie 1995). It would be an important element of the activities of the Japanese R & D laboratory as it tries to develop and source technology from overseas. This has many implications for MTI that will be explored in the next chapter.

A brief note on globalization and national innovation systems

Despite increases in globalization, the nation remains the most important innovation arena. What happens inside national borders largely determines the success of the national economic endeavour. It is still within the national arena, for example, that the majority of regulations are determined, and that policies for research, training, the protection of intellectual property, access to finance for development, and so on are decided. Furthermore, research shows the importance of national, cultural differences in approaches to product development (Ettlie et al. *1993). Archibugi and Michie (1995) confirm this analysis from another point of view. They present evidence that, although transnational firms exploit technological opportunities in a global context and collaborate internationally, they also rely heavily on home-based technological infrastructure for the generation of technology. They conclude that national innovation policy remains an important determinant of the international competitiveness of nations, despite the globalization trend.*

This reliance is partly due to the crucial importance of innovation 'speed'—the reduction of lead times for innovations to reach the market if they are to be profitable. This speed is significantly affected by conditions in the home market, which impact on the ease of creating and operating various technological developments (Teubal 1996). If the home market is not appropriate (has the wrong industrial and especially customer mix, a poor R & D base, poor provision of finance) initial marketing, sales, and product testing cannot be carried out effectively.

It is for reasons such as this that Porter (1990: 19) concludes that, 'while globalization of competition might appear to make the nation less important, instead it seems to make it more so. With fewer impediments to trade to shelter uncompetitive domestic firms and industries, the home nation takes on growing significance because it is the source of the skills and technology that underpin competitive advantage. This point is developed by Meyer-Krahmer (1999: 3), who argues that

> *the winners in a closely interlinked world economy will probably be those locations which, owing to competence and openness, become centres of information, communication and knowledge application. It is the overall attractiveness of a location which is important. Future national innovation policies will have to increase this attractiveness, not only by encouraging individual breakthroughs, but also by supporting innovative networks, while at the same time optimising a number of other locational factors in order to facilitate leading edge markets.*

CONCLUSIONS

This chapter has described some of the major changes occurring in industrial society that affect MTI. It has analysed how revolutionary changes in technology require concomitant changes in industrial structure, management, and organization to take advantage of new technological opportunities. The changes have led some analysts to argue that we are moving to a 'knowledge economy'. These changes can be profound and require firms to adjust to them rapidly. The chapter has analysed the importance of various kinds of system and discussed the value of the new forms of analytical tools that enable firms to understand their position within, or integrate themselves in, these systems and networks so as to benefit from their membership. A new paradigm of management has been outlined and its importance for the new industrial structures has been described. MTI has to reflect these changes and contribute to them in creating a more efficient and empowered form of management. The innovation process is itself changing. Using the concept of the fifth-generation innovation process, it has been argued that contemporary innovation is now dependent upon greater levels of strategic and technological integration, and

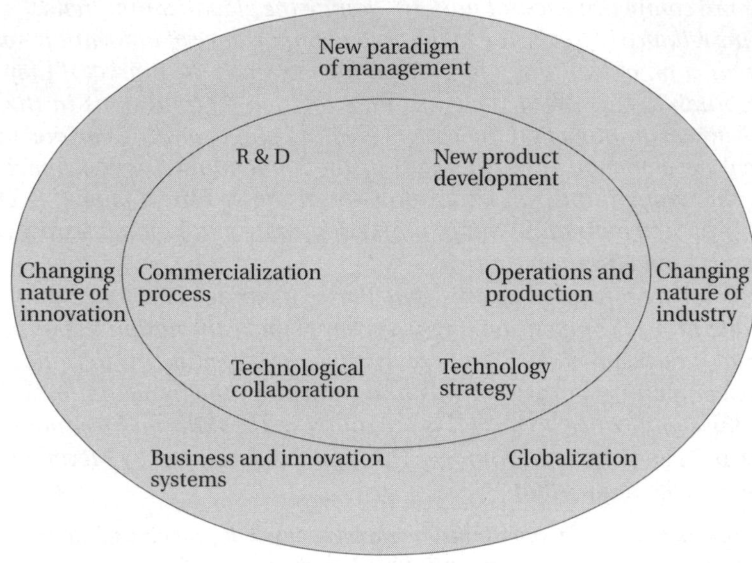

Fig. 2.5. Management of technological innovation: contextual issues

new toolkits of organizational techniques and electronic media distinguish the present innovation process from those preceding it.

The major challenges affecting MTI are illustrated in Fig. 2.5. These challenges provide the context in which MTI occurs and explain the high degree of complexity, risk technology, and innovation managers must confront.

CHAPTER

3

The Management of Research and Development

Business invests heavily in research and development (R & D). In 1997 the business sector spent over $350 billion on R & D in the OECD nations, accounting for over 69 per cent of all expenditure on R & D in these countries (OECD 1999). Some companies are huge spenders on R & D. General Motors (GM) invested over $7 billion annually in the late 1990s. Fewer than ten countries around the world spend more. Around 60 per cent of total corporate R & D in the United States now occurs in the information technology/electronics and drugs/medicine industries. Companies in these industries commonly spend over 10 per cent of their annual sales on R & D, which may constitute a significant proportion of a company's activities. For example, one in four employees of Nokia works in R & D.

This level of expenditure puts enormous pressure on companies, and managing R & D is a key element of MTI. This chapter examines the most important issues in the management of R & D, including its organizational structures, the management of R & D teams, balancing R & D portfolios, its international management, and the evaluation and assessment of research.

WHAT IS RESEARCH AND DEVELOPMENT?

Firms undertake R & D for a variety of reasons, including:

- supporting existing business activities;
- establishing new business developments;
- facilitating related business diversification;
- selling R & D services to other companies;
- providing the skills to help 'reverse engineer' competitors' products (to see how they work);
- helping predict future technological trends;
- complying with social and political expectations;
- participating in research networks;

- portraying a positive corporate image;
- creating future options through new knowledge and technology.

Also, as the research director at Microsoft explained when discussing why it had created a central research lab, 'it never hurts to have smart people around' (Cusumano and Selby 1995).

The research productivity of private-sector R & D is clearly seen in the case of AT&T's Bell Labs. Most of Bell Labs has been transferred to the company's separate manufacturing arm, Lucent Technologies, and is now focusing on shorter-term research. At its peak, however, Bell Labs had 25,000 R & D employees. It was the birthplace of the transistor, laser, solar cell, light-emitting diode, digital switching, communications satellite, electrical digital computer, cellular mobile radio, long-distance TV transmission, sound motion pictures, and stereo recording. It received more than 26,000 patents, averaging one per day since its founding in 1925.

There are various definitions of different types of research and development, none of which is completely satisfactory because of the considerable overlap between types. One common distinction is between basic and applied research. Basic research is characterized by its long-term horizons and concern with new discovery and understanding. Applied research, in contrast, is often nearer term and conducted in relatively well specified areas of enquiry. However, as Nathan Rosenberg has pointed out, basic research is often used to explain how particular technology works, and is therefore already rather prescribed. Sometimes the term 'strategic research' is used to imply longer-term research of high future potential but no immediate value to the firm conducting it. Such a description implies the ability to predict which research will have 'strategic' outcomes for the firm—an ability that is extremely rare. Other categories of research used include 'curiosity driven', 'mission oriented', and 'pacing'. Because the motives and expected outcomes of research are rarely clear-cut, the simple distinction between basic and applied research will be used here, but they must be seen as elements on a continuum where differences are not always distinct.

Although there is considerable overlap with the R & D domain, at some stage the practicality of ideas has to be demonstrated. For example, it may be possible to produce a new substance in a laboratory test tube, but can it be made in larger, potentially commercial quantities? This kind of experimental development is often undertaken in the domain known as advanced engineering. Once R & D has proved the feasibility and potential of a new product or service, later stages consist of 'pilot' operations that establish whether the product could be economically justified—that is, capable of being produced in adequate quantities and yields at acceptable prices. Design engineering occurs in the development domain when the major specifications of research or a new product project are well known.

The majority of expenditure in R & D is devoted to the development and design areas. According to the US National Science Foundation, for example,

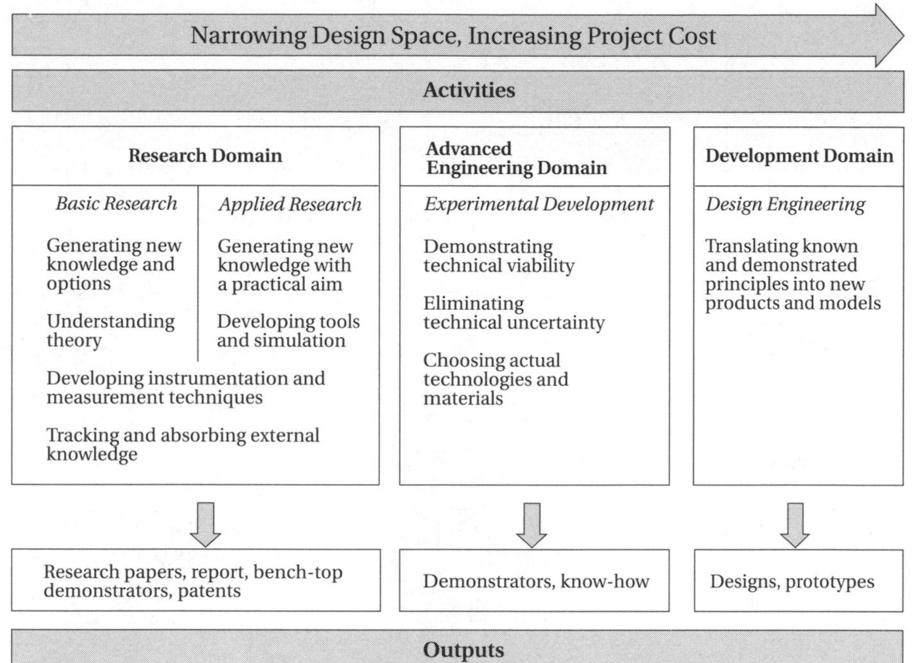

Fig. 3.1. R & D activities and outputs
Source: Arnold *et al.* (1992).

industry in the United States devotes 15 per cent of its activities to basic research, 22.5 per cent to applied research, and the remaining 62 per cent to product development (engineering and design)—proportions that have remained relatively stable since 1970 (NSF 1998).

Fig. 3.1 shows some of the major aspects of R & D. It should be emphasized that there are often overlaps and iterations, and few projects clearly follow a linear model of development between the different domains. As activities move closer towards advanced engineering and design, the opportunities to change configurations and specifications are narrowed and generally costs increase. Fig. 3.1 also shows how the outputs of the different activities vary.

CENTRALIZED AND DECENTRALIZED R & D

Whereas R & D structures in smaller firms are usually straightforward (having one site where R & D is undertaken), larger firms have choices. Firms can structure their research organizations in a number of ways. Either they can have one or more central laboratories, or they can have one or more divisional laboratories

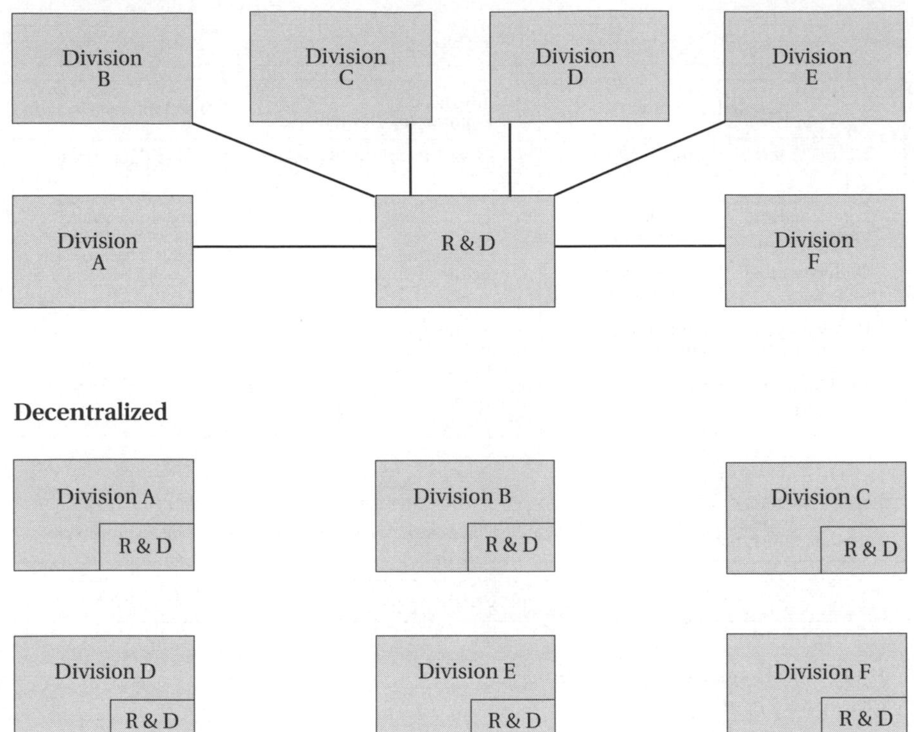

Fig. 3.2. Centralized and decentralized R & D organizational structures

reporting to business operations, or they can operate with both these forms of organization. One of the key structural questions for large, multi-divisional companies in their research activities is the extent to which they are centralized or decentralized. Fig. 3.2 shows the two extreme types of structure: completely centralized and completely decentralized. A number of issues affect companies' decisions about the extent to which R & D is centralized or decentralized.

Type of activity undertaken

When the research is more basic and likely to have longer-term implications for the firm, there are advantages in having strong links with corporate head-quarters, where longer-term strategic business decisions are made and where speculative research can be 'protected' from the more immediate demands of business divisions. When the research is more applied or advanced engineering is being performed, there are advantages in decentralized links with divisions, which are closer to the customer and can better respond to their requirements. Once general principles are established, detailed development and design activities are often best linked to particular factories. The location of basic research in the United States and Japan tends to be determined by issues such as good living environments, transportation, and the general availability of engineers. Japanese firms tend to co-locate applied research, production engineering, and manufacturing more than US and European firms (Kenney and Florida 1994).

Scale of research

Centralization of research is attractive when there is the need to achieve economies of scale—for example, when it necessary to create a 'critical mass' of researchers for a particular project, or to utilize expensive research equipment. When communication between research groups is important, there are advantages in co-location in central R & D.

Need for functional integration

When it is important to have strong links with other functions, particularly with manufacturing and marketing, then decentralization of R & D has advantages.

Recruitment/labour and other cost considerations

Decisions about R & D structures can be influenced by locational and managerial efficiency questions. So, for example, certain regions may be famous for their researchers in a particular field, or be closely located to important customers, and this may affect companies' decisions about locating research (Silicon Valley would be a classic example). Specific considerations of the needs of key individuals, such as the ability to access the expertise of particular scientists, or effectively to utilize the limited availability of highly skilled research managers, may also affect decisions about the location of R & D.

Fig. 3.3 shows the R & D organizational structure of Hitachi, the world's third largest electronics company, which has one of the most centralized R & D structures. In 1997 Hitachi spent nearly $4 billion on R & D, accounting for over 6 per cent of sales. It had thirty-seven R & D laboratories, nine of which were centralized. The central labs spent 24 per cent of Hitachi's total research budget, of which 8 per cent was on basic research and 16 per cent on applied research. It had 13,500 R & D staff, of which 9,000 worked on product development in various divisions and factories. The remaining 4,500 worked in the

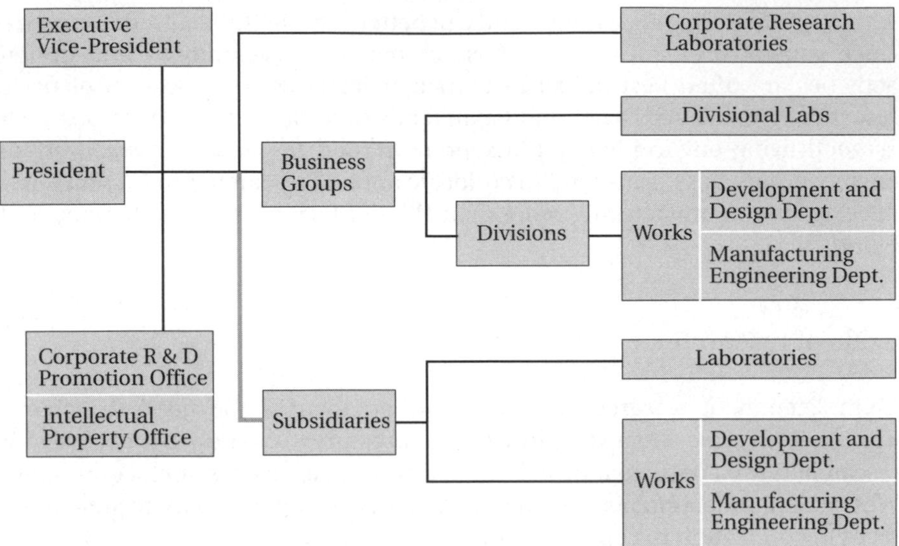

Fig. 3.3. Hitachi's R & D organization in 1997

corporate research laboratories, including the Central Research Laboratory (which had 1,300 employees mainly researching electronics) and the Advanced Research Laboratory (170 staff working on fundamental science).

According to Sigurdson (1998),

> the funding pattern for R & D in Hitachi appears to be different in one major aspect compared with similar companies in the USA and Europe in that the corporate laboratories play a more important role . . . Commissioned research in the corporate research laboratories—coming from business divisions, subsidiary companies or contacts from other laboratories—constitute 70 per cent of the R & D activities. Commissioned research in the Central Research Laboratory is 55 per cent and nil in the Hitachi Advanced Research Laboratory where researchers only do fundamental science.

Sigurdson also argues that Hitachi may be following the pattern of other major companies, such as Siemens and General Electric (GE), by steadily decentralizing its R & D. This pattern may also be occurring in other Japanese firms. Toshiba, for example, began its transfer of R & D to the business divisions some years ago. Its R & D centre's staffing was reduced from 2,000 to 1,600 in 1998. In part this reflects the financial problems faced by many Japanese firms in the late 1990s, but also the need to transfer research as soon as possible to businesses.

Despite this trend towards decentralized R & D, centralized R & D can be very important. Roberts's (1994) study of ninety-five major international companies showed that central corporate research organization is the primary

Box 3.1. ***Overall rank-ordered importance of sources for research and development work***

Rank	Research	Development
1	Central corporate research	Internal R & D within division
2	Internal R & D within division	Joint ventures/alliances
3	Sponsored university research	Central corporate research
4	Recruiting students	Incorporating supplier technology
5	University liaison programmes	Licensing
6	Consultants/contract R & D	Acquisition of external technologies
7	Continuing education	Acquisition of products
8	Joint ventures/alliances	Consultants/contract R & D

Source: Roberts (1994).

source of technological information and advance in Japan, Europe, and the United States in most industries (Box 3.1).

In the case of GE, its central R & D centre employs only about 10 per cent of GE scientists and engineers, but generates about 30 per cent of the company's patents and produces one-third of its technical publications. In 1996 the R & D centre's staff was awarded 325 of the company's 994 patents, and they wrote approximately 200 journal articles, book chapters, and conference proceedings.

A common difficulty facing R & D directors is justifying the funding for centralized R & D. Explaining some of the reasons for R & D described earlier, such as social and political expectations and future values and options, usually holds little persuasion for divisional business managers concerned about their next quarter's results. Possible types of funding arrangement include the use of a *fixed rate*: for example, all divisions pay an agreed percentage of sales to central R & D. Another mechanism uses *competitive* systems, whereby headquarters identifies a number of research projects, and central and decentralized labs compete to win the projects. Other methods use a simple *contract* system, whereby the central lab undertakes research for the divisions or headquarters on a contract-by-contract or commission basis. Another approach uses a *levy* on divisions (if they are especially profitable one year, for example) to pay for particular projects. In practice, combinations of these funding mechanisms are used. Hitachi, for example, funds its central labs both by direct grant and from commissioned projects from the business divisions (an exception is its Advanced Research Laboratory, which undertakes only basic research and does no commissioned work).

Both centralized and decentralized R & D have advantages and disadvantages, and the ability to fund both is often unaffordable to all but the wealthiest corporations, like Hitachi. In his excellent study of the structure of R & D in industry Rubinstein (1989) found that companies tended to undergo phases of centralization then decentralization then centralization in their research, and they exhibited a high degree of dissatisfaction with their organization of R & D in whatever form it took. Roberts notes that US firms are becoming much more decentralized in their R & D than European and Japanese firms.

> I believe that these control changes take place in R & D organizations in cycles, especially for research, with about seven-to-ten years for the half-cycle. In the US, my opinion is we are nearing the end of the half-cycle of decentralization of R & D control, i.e. moving control of budgets and programs down to the divisional or business-unit level. I expect that within a few years US companies will start to recentralize control of R & D as they find the problems of technological blindsiding and short-term investment management begin to dominate competitive issues at the business-unit level. American firms will again begin to make long-term investments, perhaps by creating corporate centers of excellence in areas of core technology, putting more money into longer-term corporate research. (Roberts 1994: 9)

There are advantages and disadvantages to all forms of R & D organization, and the structures chosen will vary over time. Some companies, such as Lucent and 3M, co-locate their divisional and central labs in order to try to gain the benefits of both forms. Ultimately, the R & D organization should reflect the strategy pursued by the company. This, and the question of the coordination of central corporate R & D with divisional activities, will be examined in Chapter 7.

MANAGING RESEARCH TEAMS

Research management has to be sensitive to the aims and needs of different kinds of R & D activities. Firms have different control structures for different kinds of research activity. The classic study of innovation management in the 1960s showed how firms were organized in 'organic' or 'mechanistic' forms according to their aims (Burns and Stalker 1961). Organic forms are associated with the encouragement of flexibility and initiative and the avoidance of prescriptive communications channels and authority. They have a community structure of control and devolved technical authority. Mechanistic forms are associated with hierarchical control, authority and communications, demand for obedience, and very precise definitions of methods. These are two ideal types of organization, but subsequent research has shown the advantages for innovation of organic forms, where there is a high degree of technological complexity. Management organization also reflects the need, when projects are more easily predictable, specified, and controllable, to move from 'loose' to 'tight' management control. The management of R & D is a prime case of the need for organic organization, although it is conceivable that more mechanistic approaches can be used closer to the engineering and design domains, in the unlikely event of the firm seeing little value in encouraging creativity in these areas.

There is a substantial literature on the management of research teams and projects. The research on teams is often large scale, involving many hundreds if not thousands of respondents. It provides fascinating insights into a number of areas. For example, the productivity of research teams declines after about five years, as team members begin to restrict their external networks, and research productivity declines for researchers in their early to mid-30s and then intriguingly increases again in their 60s (Katz and Allen 1982; Farris *et al.* 1995). In contrast to this latter finding, other research shows that there are research 'stars' whose research output increases with age (Goldberg and Shenhav, 1984).

Within research teams, individuals have different approaches towards communicating technological information. Scientists in the team usually rely on primary literature (journals), are outward looking, unconstrained by the firm, and relate to other scientists as peers. Engineers, on the other hand, tend to rely on secondary literature (trade and professional magazines), are inward looking, are concerned only with problems within the firm, and relate to colleagues within the firm as their peers. Effective research-team management enables both types of researcher to operate to the advantage of the firm.

There is a wide range of 'stage-gate' project management systems used to filter and evaluate research activities in a rational and impartial manner (see Chapter 4). The use of these techniques, importantly, is also seen to be fair by those affected by the decisions that these systems help management reach. Box 3.2 shows a management system based on the roles of key individuals during the course of a project's development. As projects develop, evaluation techniques are used at the various stages before allowing progression to the next level. It was successfully used by Celltech, the UK biotechnology firm.

Developments in groupware during the 1990s, such as that mandated for

Box 3.2. ***Project management system based on the roles of key individuals***

Stage	Person responsible for each stage
Idea	(Self Appointed) idea champion*
Candidate	Candidate champion*
Research project	Research project manager
Development project	Development project manager
Product	Product manager

* Part-time responsibilities

large-scale project management in the US Department of Defense, can democratize the research process by making inputs into problem solving exercises anonymous so that ideas are evaluated according to their usefulness rather than their source (which can be helpful in hierarchical organizations).

A US study of project team management found a range of factors that enhance and impede project performance (Box 3.3). They found that the range of factors that encourage team effectiveness is less 'instrumental' (involving relatively simple, tangible issues) and more 'subjective' (for example, to do with professional satisfaction and recognition). The factors that inhibit project performance are, however, much more directly objective and are related to practical issues of project management, such as project objectives and resource limitations.

Research into the effectiveness of research teams in Japan (in companies such as Nissan) shows the value of integrated, interdisciplinary teams (Graves 1994). Interdisciplinary or cross-functional teams involve representatives of a wide range of functions within the firm and bring benefits for innovation (Lutz 1994). A study of mainframe computer companies (AT&T, Bull, Fujitsu, Hitachi, IBM, ICL, Mitsubishi Electric, NEC, Toshiba, and Unisys) also shows the value of creating integrated R & D teams with a 'systems focus' (Iansiti 1993). It was found that systems-focused companies achieve the best product in the shortest time and at the lowest cost. The integrated R & D team included a core group of managers representing both the

Box 3.3. ***Factors enhancing and impeding project performance***

Rank	Drivers	Barriers
1	Professionally stimulating and challenging work	Unclear project objectives
2	Sense of accomplishment	Insufficient resources
3	Proper experience and skills of managers	Role conflict and power struggle amongst team
4	Good overall direction and leadership	Lack of commitment to team
5	Proper experience and skills of other staff	Poor job security
6	Professional growth potential	Excessive changes of project scope, specification, schedule and budget

Source: Thamhain and Wilemon (1987).

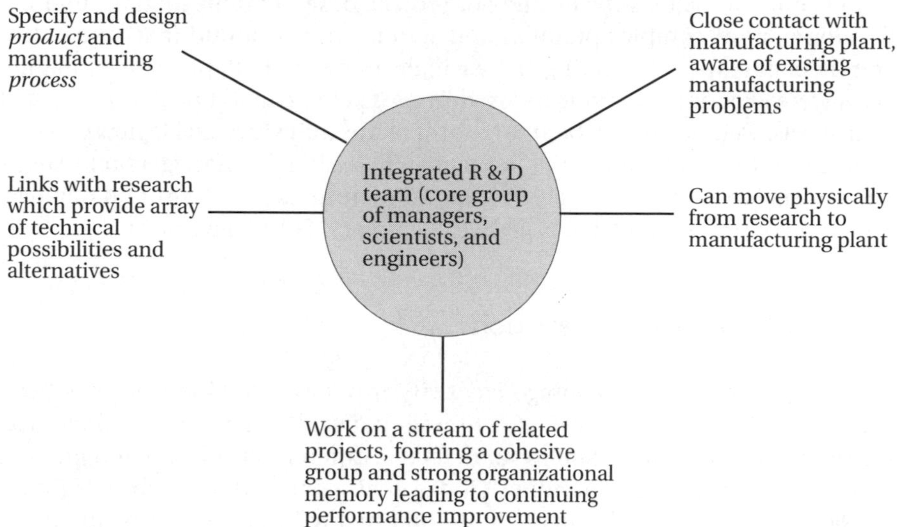

Fig. 3.4. Integrated R & D teams
Source: Iansiti (1993).

research function and existing manufacturing capabilities. Fig. 3.4 shows some of the characteristics of integrated R & D teams.

A similar approach has proved very effective for US car manufacturer Chrysler. According to Lester (1998: 75), the Chrysler approach involved

> taking the concept of product team autonomy much further than Ford or GM. The company created five autonomous 'platform teams', each devoted to a different product line (small cars, large cars, minivans, Jeeps and trucks). Each team consisted of all the people needed to design and produce a new car, including manufacturing, purchasing, finance and marketing professionals, hourly manufacturing workers, and even representatives of key outside suppliers, as well as engineering and design staff . . .
>
> Hardly any bureaucracy has since been allowed to remain above the platform teams, either. Each team is presided over by a top executive, giving it a voice at the highest levels of the corporation. The teams receive precise instructions from the top regarding key vehicle specifications—engine power, weight, fuel economy, and so on—as well as the total budgets for the projects. Once these specifications are met, however, the teams are free to work out how to meet them, with little or no interference from senior management. In return for this increased authority, the platform teams are also held accountable for their performance. And since the members of the team are judged by the overall success of the vehicle, rather than their particular bit of it, they are inclined to find ways to cooperate with each other.

The use of these platform teams is one of the major reasons, according to Lester, for the regeneration of Chrysler in the 1990s.

In Gann and Salter's (1998) study of project-based organization around the development of complex products and systems, it was found that 'integrative competences' were becoming increasingly important. Rapid team-building skills were described by project-based firms as core capabilities for personnel at all levels. 'People need to be able to form teams quickly to tackle new projects or respond to events in existing projects. Professionals, managers and shop-floor operatives need to be able to respond to unforeseen events and deploy a high level of problem solving expertise' (Gann and Salter 1998: 443).

Managing creativity in research

Managing research to encourage creativity involves resolving what has been called 'the dialectical process of synthesis between multiple dilemmas (e.g. freedom and control, flexibility and focus, differentiation and integration, incrementalism and discontinuity)' (Judge *et al.* 1997). In their study of eight US biotechnology firms, Judge *et al.* found that highly innovative firms managed these dilemmas and created 'focused communities', whose innovativeness was encouraged by features such as family atmospheres, trust, caring, contemplation, and self-examination.

Some companies are remarkable for the efforts they put into encouraging creativity in research. During the late 1980s the small UK biotechnology company Celltech had a scientific advisory group of international eminence for assessing scientific performance and providing external 'peer review' at the highest level. The company held over fifty internal and external scientific seminars a year, the equivalent of many leading university departments.

In many creative firms, the researchers themselves have high levels of discretion over the selection and conduct of research projects. Dow Chemical's chemists in their central R & D organization have responsibility for planning, implementing, and evaluating their own work on projects. Celltech allowed researchers to spend 10 per cent of their time on personal research agendas. This mode of organizing creativity is not the preserve of smaller, high-technology companies. Nippon Steel, the largest steel manufacturer in the world, also allows its 1,200 R & D staff 10 per cent of their time to spend on projects of their own initiative. 3M allows all its employees, not only R & D staff, 15 per cent of their time to work on their own projects. Celltech's buildings were designed to maximize the amount of communication between researchers, with numbers of large, open meeting rooms. The level of resources available to researchers and the freedom to undertake research of their own help explain the movement of very talented researchers in molecular biology from the public sector to such companies in the private sector in the United States. Fuji Xerox's Palo Alto California laboratory uses many of the most common methods of encouraging communications, such as roundtables on particular areas of interest, talks by eminent visitors, weekly work-here-in-progress seminars by employees, weekly lunch meetings, and an internship programme.

Judge *et al.* (1997) measured 'innovativeness' by means of citations analysis—measuring the extent to which academic papers and patents are referenced in subsequent papers and patents—on the theory that, the more they are cited, the better their quality. They determined that the most innovative firms had the following characteristics.

- Researchers possessed 'operational' autonomy, inasmuch as they acted entrepreneurially and had a sense of individual accomplishment. And top managers possessed 'strategic' autonomy, which aligned individual researcher interests with organizational aims. Too little control by top management was found to cause a disconnection between business goals and research enterprise. A balance between operational and strategic autonomy promoted innovation by encouraging researchers to be creative in organizationally beneficial ways.
- Individual and group success was rewarded extensively through 'intrinsic' methods of personal acknowledgement and recognition by managers and peers, rather than simple 'extrinsic' impersonal rewards of salary increases, bonuses, and stock options.
- There was evidence of 'group cohesiveness', developed through particular attention being paid to recruiting people whose 'faces fitted' and who slotted into the social environment.
- Established goals were reasonable and deadlines were flexible.
- The firm possessed 'organizational slack', which allowed adaptation in strategies pursued in the past and the present, and projected into the future.

Effective group and team management is essential for creativity and the success of R & D projects, but it is also important to manage key individuals. In their studies of the productivity of research labs in the United States and Japan, Narin and Beitzman (1995) identified the importance of key employees. It is commonly one or two 'stars' who are responsible for the labs' productivity. It is the job of research managers to identify these people, nurture their creativity, and ensure they remain with the company.

The challenges of encouraging creativity in R & D is seen clearly in Samsung Electronics Corporation (SEC), part of the Samsung Group. SEC employed 11,000 in R & D and spent 12 per cent of sales on R & D in 1995. It is the largest manufacturer of DRAM (dynamic random access memory) and memory chips in the world. It has eight domestic and thirteen international R & D centres. Samsung has been extraordinarily adept at catching up with world-best science and technology. The challenge it faced in the late 1990s was to lead the development of science and technology, and this requires creativity—something Korean firms have not been noted for in the past.

Researchers in Samsung have faced the following problems in the past: poor project management (control and evaluation is undertaken by non-specialists); high levels of bureaucracy and rigid procedures (for example, in requiring excessive documentation); slow information processing systems in areas such

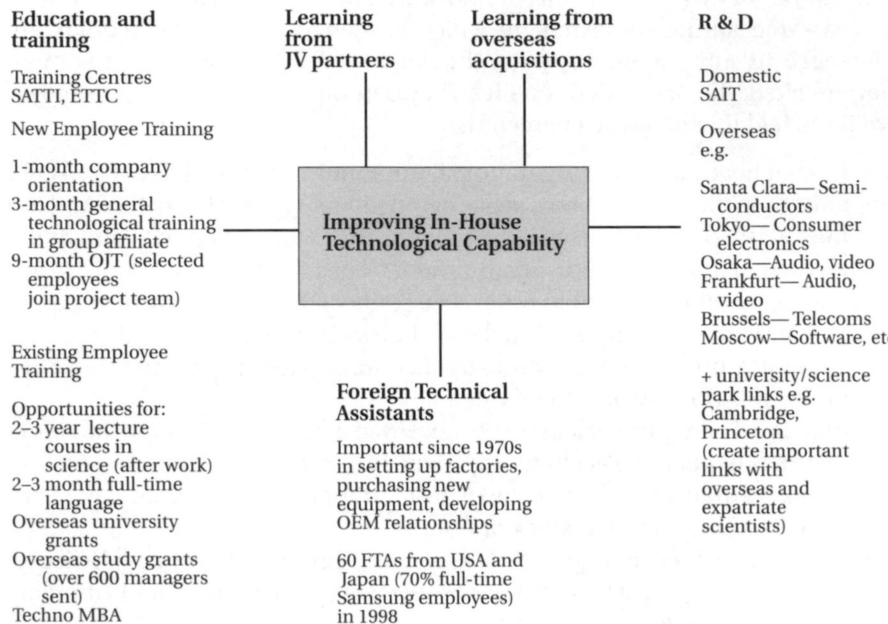

Fig. 3.5. SEC's sources of technological learning
Source: based on Dodgson and Kim (1997).

as patent searches; lack of involvement of personnel from the manufacturing function; too high a workload (working on too many projects); and lack of funding for longer-term research. Major changes are under way in Samsung to overcome these constraints on creativity, including a range of training courses in MTI. One significant stimulus of change in Samsung has been learning from international experiences. This will be examined in later chapters, but Fig. 3.5 illustrates the ways Samsung Electronics Corporation is learning about technology. The combination of all these efforts is working. In 1998 it was the fourth largest patentor in the United States (CHI Research 1999).

A brief note on technological gatekeepers

One important role in research teams is that of the 'technological gatekeeper' (Allen 1977). This is an identifiable individual who is a good transceiver of information. He or she

- *brings information into the firm/team;*
- *disseminates information throughout the firm/team;*
- *attends conferences, reads journals, and talks avidly with fellows;*
- *must be encouraged and rewarded.*

His or her presence in an R & D project is correlated with finding superior technical solutions. More recent research shows the role of the technological gatekeeper continues to be important (Macdonald and Williams 1994).

BALANCING RESEARCH PORTFOLIOS

Perhaps the greatest problem of research management is the tendency of companies towards short-termism. Few industrialists have the foresight of Konosuke Matsushita, who founded his company in 1932 with a 250 year plan! (Barnet and Cavanagh 1994). Most 'long-term' industrial R & D has a time horizon of five years. However, many large Japanese firms' R & D horizons extend to 10–15 years, and 3M's central research has a timescale of over ten years. Pharmaceutical firms, albeit with the confidence of potentially solid patent protection, can expect the production of a new drug to take up to fourteen years between initial screening and marketing. In 1998 IBM began to introduce its first chips coated with copper, rather than aluminium, following fifteen years of research (*The Economist*, 6 June 1998).

Industrial firms are conscious of the high degree of 'failure' in research, knowing it to be a fundamentally uncertain and unpredictable process. Surveys of the introduction of new products show that less than one in ten new product ideas succeeds in the market (see Chapter 4). This tolerance, although not easily accepted (as most R & D directors will testify), holds many lessons for those trying to manage research. Success and failure are essentially unpredictable, serendipity and luck are important, and some failure is inevitable and cannot be prevented.

R & D directors face a dilemma in organizing their budgets to meet both long- and short-term objectives. Whereas some research shows that R & D is likely to be more successful when it is related to firms' existing technology bases, these projects ignore new developments that can potentially provide future competitive advantages. This is a concern recognized by many US R & D managers, who feel their portfolios are becoming too short term (Roberts 1995).

There are advantages to research managers operating their project research portfolios the way funds managers' operate financial portfolios. There are benefits in balancing short-term, low-risk, low-reward investments with longer-term, high-risk, high-reward investments. The dilemma is perfectly encapsulated by Jack Welch, CEO of GE, who argues: 'Anyone can manage short. Anyone can manage long. Balancing those two things is what management is' (*Business Week*, 8 June 1998). Decisions on the balance, focus, and sustainability of R & D portfolios should depend on the company's strategy. This is examined in Chapter 6.

Fig. 3.6 characterizes a potential portfolio mix. Research managers need to balance risks with the potential rewards from R & D investments, in line with the

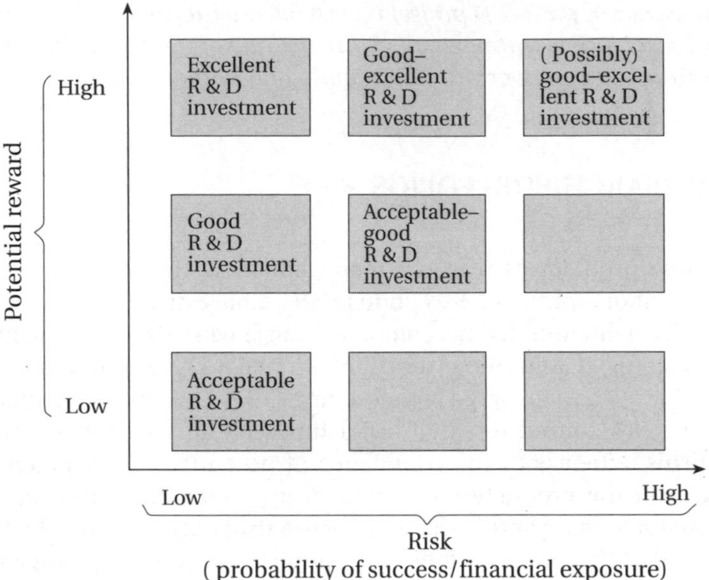

Fig. 3.6. Research portfolio mix
Source: Roussel *et al.* (1991).

technology strategy of the firm and the level of resources—finance, personnel, and equipment—available.

Research by Henderson (1994) on 120 pharmaceutical development programmes over thirty years showed how portfolio diversity was the key to success. Examining the mean number of patents per programme, she found the highest productivity where there were between six and ten programmes. Productivity increased from one to six or seven programmes, and gradually decreased after nine or ten. Henderson concludes that 'the most productive companies are not only diverse enough to enable them to leverage their specialized scientific expertise but also related sufficiently to allow them to benefit from the cross-fertilization of ideas'.

A brief note on the importance of basic research

The ability to conduct and access basic research will increasingly determine competitive advantage in high-technology industries. This is seen in the case-study companies in Chapter 1. The US biotechnology firm emerged from the science base. The Japanese R & D company depends upon its links with universities around the world. The Taiwanese machine tool company needs access to basic research to get it to the next level of its technological develop-

ment. There is a close link between basic research and industrial application (Grupp 1994; Meyer-Krahmer 1997). Many firms are already significant performers of basic research, an area that has traditionally been seen as the preserve of universities and large government laboratories. For example, IBM discovered superconductivity and Bell Labs discovered radio static (leading to the development of radio astronomy). Bell Labs has won seven Nobel Prizes for physics. The scale of corporate basic research is seen in the way companies are increasingly publishing in the scientific literature. Companies such as Philips, Hitachi, ICI, Siemens, Hoechst, and Toshiba publish as much as good middle-sized international universities like Sussex in England and Keio in Japan (Hicks 1995). In the biological and physical sciences and in electrical engineering, companies produce scientific papers cited as much or more than those from top US universities (the more papers are cited in other publications, the more they are considered to be influential) (Hicks 1995). GM employs 360 Ph.D.s in its R & D Center in Detroit, GE employs 500 Ph.D.s in the GE Research and Development Center, and Bell Labs, at its peak, employed nearly 4,000 Ph.D.s.

Why do firms do so much basic research? As Rosenberg (1990) notes, firms do basic research for a variety of reasons, including sending signals to potential employees and collaborators that the firm is serious about and competent in its areas of science. Pavitt (1993) argues that the economic contribution of basic research lies in the way it trains researchers, enables membership of (international) networks, develops new research techniques and instrumentation, and creates a store of 'background knowledge' that improves the effectiveness of research activities and the ability to solve large, complex problems. Basic research in firms is as much about creating future options (or option value) for future problem solving and participation in research communities as it is about tangible scientific outcomes. An example is provided by Toshiba and Hitachi, who are both undertaking basic research with Cambridge University into quantum physics. They see this as providing the long-term potential for replacing existing integrated circuit manufacturing processes. Japanese companies, such as Hitachi and Mitsubishi, also undertake basic research to improve their corporate image and make recruiting easier, even when there is no immediate commercial pay-off (Kenney and Florida 1994).

Another reason for undertaking basic research is serendipity: the possibility of learning useful things unexpectedly when you are actually looking for something else. The outcomes of scientific enquiry can never be predicted accurately. One of the major scientific breakthroughs underpinning the development of optoelectronics, for example, derived from a theoretical physicist's work researching the nature of light on the eyes of bees.

Basic research in academia is also important for industrial firms in many industries. Mansfield's research has shown how in two periods, 1975–85 and 1986–94, in a wide range of industries, over 10 per cent of the new products and processes introduced in these industries could not have been developed (without substantial delay) in the absence of recent academic research. He estimates *that his sample of seventy-seven large firms produced sales of $44 billion from products, and savings of $17 billion due to new processes, first commercialized in the 1991–4 period and based on recent academic research (Mansfield 1998). Furthermore, some of the scientific instruments developed as tools for basic research in academia have had profound impacts on industry. The computer, laser, and the Internet provide three important examples.*

Basic research is also a major source of intellectual property rights. Intellectual property rights will be discussed in Chapter 8, but here it is sufficient to note that they are an important output from R & D investments, an essential aspect of firms' ability to appropriate return from these investments, and they need to be carefully managed.

While basic research is important, it is highly unpredictable and its long-term horizons conflict with the usual short-term financial pressures firms face.

MANAGING INTERNATIONAL R & D

Why internationalize R & D?

Managing R & D in home countries is complex and demanding, but managing it internationally adds considerably to these difficulties. Many firms see R & D as such an important core asset that they are wary of losing control over it by conducting it in bases overseas. Aware of the importance of good communications for the management of R & D, they are concerned that distance will raise barriers to the free flow of information and knowledge. Yet many firms internationalize their R & D, and they do so for a number of reasons. The relative importance, and balance, of these objectives will vary between countries and industrial sectors, and each has implications for MTI. Internationalization allows for the following.

- *Proximity to market and customers.* R & D can develop and design products appropriate to local conditions, demand, standards, and regulations.
- *Support for local manufacturing.* Local R & D can quickly assist in overcoming manufacturing problems experienced locally.
- *Responses to political factors/government policies.* Some governments require overseas companies investing in production plants in their countries to undertake R & D there as well. Some governments run 'offset' schemes, where access to government business depends on certain levels of local R & D activity.
- *Exploitation of foreign R & D resources.* The availability of R & D expertise may attract overseas firms. High-quality and low-cost software engineers in India and Russia, for example, have attracted significant overseas corporate investment. Similarly, Hitachi has research laboratories in Cambridge, UK, Dublin, Milan, and Sophia Antipolis, France. It has a

semiconductor research lab in California, an automotive products lab in Detroit, and an advanced TV and systems lab associated with Princeton University (which also has co-located labs from Siemens, NEC, Matsushita, and Toshiba). The aims of international investment vary with the nature of the technology being used and developed. A study of international R & D investments in India showed that the major motives for investing in R & D were proximity to manufacturing for conventional technologies, and availability of R & D personnel for new technologies (P. Reddy 1997).

- *Parallel development.* Some companies use competing teams with different approaches to solving a particular project. When these teams are international, there is likely to be greater diversity in the possible solutions considered, and a higher level of competition.
- *Specialization strategies of individual subsidiaries.* Multinational companies may decide to use particular countries as the global base for certain product families, and these will require R & D support.
- *Multiple learning inputs.* The knowledge and expertise accumulated in laboratories around the world may differ from and be greater in combination than that found within individual countries. General Motors in the USA, for example, is working with designers and engineers from Opel, Vauxhall, Saab, and Holden to share resources and best practices to develop vehicles to suit different customer needs using fewer vehicle platforms (Baker 1997).
- *Network forming.* International R & D allow firms to create the all-important personal networks with researchers in international universities and other companies. Nortel, for example, claims to work with 100 academic institutions, and Siemens claims to work with over 200 professors.

Gerybadze and Reger (1998: 194) argue that there are four key determinants driving the internationalization of R & D.

- The relative importance of knowledge generation as a constituent of value-added in the corporation. In their study of twenty-one multinationals, they found that 20–40 per cent of value-added could be attributed to knowledge-generating activities.
- The distinction of highly advanced scientific and technological knowledge bases, for which 'centres of gravity' and critical assets are often strongly concentrated in unique locations.
- The global distribution of customer-related knowledge pools and of country-specific conditions for learning from advanced users.
- The opportunities and restraints for decomposing knowledge-generating activities across locations. This depends on the industry as well as product-specific characteristics of interaction between R & D, manufacturing, and marketing, and on the degree of modularity or decomposability between functions.

In some cases, these factors lead to the development of 'centres of competence' with global responsibilities for particular technologies or products, outside the multinational's home country (European Commission 1998). According to Gerybadze and Reger (1999), the complexity of previous forms of organization has been rationalized such that many leading companies have adopted a strategy of multiple centres of learning with one dominant centre of coordination.

The extent of internationalized R & D

Firms have utilized international R & D for some years. Cantwell (1998) found that large European and US firms carried out around 7 per cent of their R & D overseas in the 1930s. For the reasons listed above, it is argued that firms are increasingly undertaking international R & D. ABB, a quintessential globalized company, spends 90 per cent of its R & D overseas, and major companies such as Philips, IBM, and Hoechst spend over 50 per cent of their R & D internationally.[1] Box 3.4 and Fig. 3.7 show the distribution of overseas R & D in selected companies.

Managing and organizing international R & D

In one of the most comprehensive studies of globalized R & D, Florida (1997) examined foreign affiliated R & D laboratories in the United States. He found that

- foreign firms invested in R & D in the United States to gain access to science and technology and to network with the science and technology community;
- rather than investing in R & D in order to support local manufacturing or to adapt products to local market needs, or to act as 'listening posts' picking up information, foreign firms undertook R & D in order to innovate new products;
- there were broad differences in the motivation for investing in R & D, with biotechnology investments, for example, being made primarily as a means of linking with the US science base, whilst in automotives the purpose lay in assisting production plants;
- foreign-owned R & D affiliates possessed considerable autonomy in the management of their activities, in determining research priorities and employee recruitment.

[1] For a detailed discussion of the extent and nature of international R & D, see the special edition of *Research Policy,* 28/2–3 (1999).

Box 3.4. ***Overseas research centres of selected companies***

Company	Total number of research centres	Number of international research centres
Nortel	41	17
Lucent Technologies	41	16
Nokia	36	11
Dow Chemicals	19	14
IBM	8	5
Hewlett-Packard	5	3
Xerox	5	2

Note: These data are for numbers of centres, and do not reflect their size or importance.

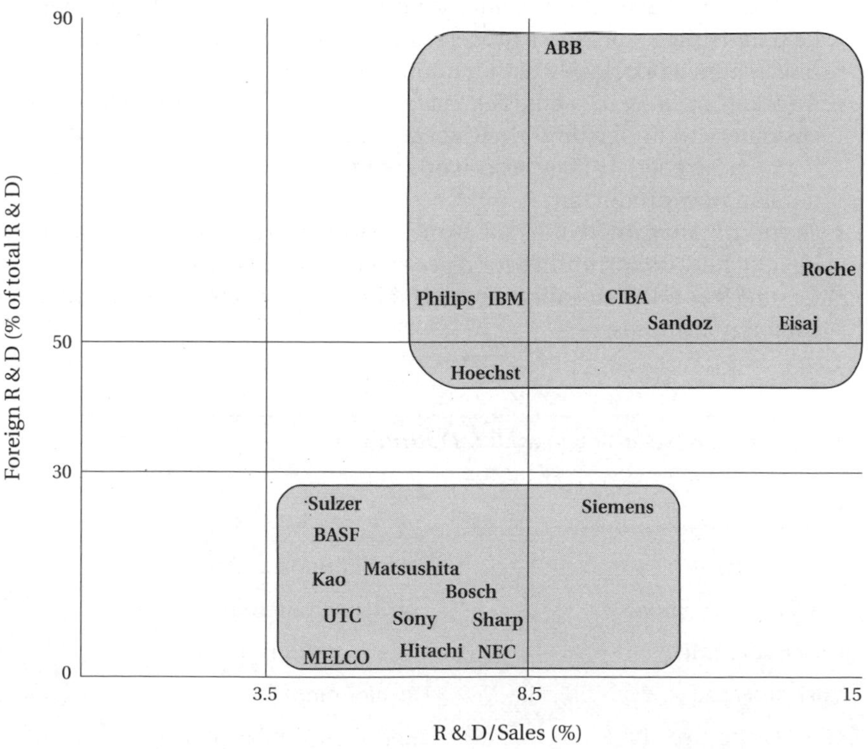

Fig. 3.7. R & D intensity of overseas R & D, selected major international companies
Source: Gerybadze and Reger (1997).

There is great variety in the extent to which firms' R & D is internationalized, with many consequences for management. The Carnegie Bosch Institute, for example, distinguishes international R & D strategies that are 'regional' or 'global' (Box 3.5). A global strategy has a product whose development draws upon expertise from around the world. An example of globalized R & D is General Motors' efforts to build more vehicles from common platforms using common components. Alternatively, a regional strategy will produce differentiated products for a limited number of different markets, using expertise only from those particular markets. An example would be a pharmaceutical company that targets the French and British markets with different requirements for drug delivery (the British tend to prefer pills, the French, injections).

Dow Chemicals provides an example of a firm that organizes its $800 million plus research activities according to both function and geographical area of activity. It has the following categories.

- *Product R & D* activities focus on particular products, such as ceramics, polymers, and coatings. Each global business within Dow has associated laboratories whose aim is to develop new products and improve existing ones.
- *Process R & D* activities include computer simulations and pilot plants to help transform laboratory processes into commercial operations. Process researchers work closely with manufacturing personnel.
- *Application R & D* activities have researchers working closely with customers to understand their needs and match these needs with Dow's research strengths. They also work with Dow's marketing staff to help develop new products.
- *Technical Service* activities focus on customer problem solving, and help identify new opportunities for researchers elsewhere in Dow.
- *Central R & D* has the aim of undertaking new research and creating new company ventures.

Box 3.5. ***Global versus regional R & D strategies***

Global	Regional
Global market	Several important markets
Similar requirements	Diverse requirements
Cost leadership	Differentiation
Basic research	Development
Global expertise	Special regional expertise
Many locations	Few locations

Source: Carnegie Bosch Institute (1995).

- *Leveraged Technologies R & D* are also known as Global Core Technologies and are organized into four major capability areas: *analytical sciences* provide measurements expertise; *complex molecule process R & D* assist the development of manufacturing processes for complex organic molecules; *computing, modelling, and information sciences* provide expertise to increase the speed and decrease the cost of new product development and research; and *engineering sciences/market development* assist clients with chemical engineering expertise.

In Dow Chemicals an international coordination element is introduced into the issues of organizing R & D so that both the needs of the market are met and longer-term opportunities are not ignored (see the earlier discussion on centralization/decentralization).

The management requirements depend also on whether R & D is 'home-base augmenting' or 'home-base-exploiting' (Kuemmerle 1997). As the titles suggest, home-base exploiting R & D is primarily concerned with using the R & D of the home base and adapting it to local requirements. Home-base augmenting requires links into the foreign R & D systems so as to add extra knowledge to that which exists in the home base (Fig. 3.8).

A number of management styles and planning and control systems are used in international R & D. These can vary from what is called 'absolute centralization', where the overseas lab does exactly what it is told, to 'total freedom', where it does what it likes (Behrman and Fischer 1980). Both of these are rare,

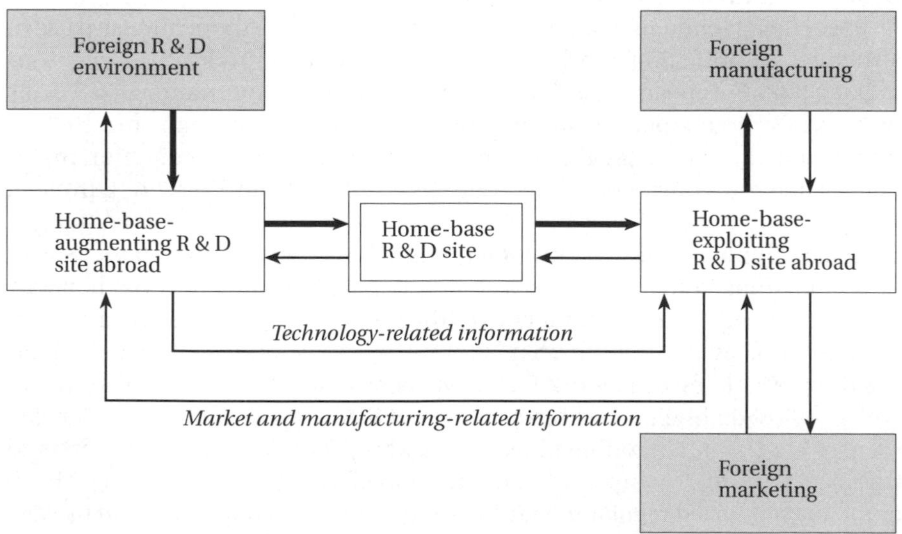

Fig. 3.8. How information flows between home-base and foreign R & D sites
Source: Kuemmerle (1997).

and more commonly management styles vary between 'participative centralization' and 'supervised freedom'. The actual choice is dynamic and changes over time. It is influenced by the characteristics of the technology and the nature of the company's culture (De Meyer 1993). Florida's (1997) study found that foreign firms did not impose their management styles upon US R & D organizations, but allowed them to follow US practices, enabling the firms to learn about the management of R & D.

Reger (1997) discusses a number of mechanisms for coordinating international R & D, including: *structural and formal mechanisms* (the use of policies and plans, coordinating bodies, and centralized/decentralized structures), *hybrid/overlaying mechanisms* (such as strategic or core projects and interfunctional teams), *informal mechanisms* (interpersonal interchange and personal development), and *internal markets* (internal contracts and research contracts) (Fig. 3.9).

One of the most important methods of coordinating the international management of R & D is a technology steering committee (Kuemmerle 1997; Reger 1997). Members of such committees need to be sufficiently senior to be able to mobilize resources at short notice and they need to be actively involved in the management and supervision of R & D programmes. In Kuemmerle's study, members included the heads of major existing R & D sites. In a study of forty-five of the largest pharmaceutical companies around the world, it was found that thirty-two operated an international R & D committee. These committees had an average membership of fifteen. Twenty-five of the thirty-two made decisions on project selection, and sixteen made decisions about financial allocations (Halliday *et al.* 1997).

Reger (1997) found that Japanese firms tended to make more intensive use of structural coordination mechanisms and be more centralized in decision-making than European firms. Many of the most highly internationalized firms in his study, both Japanese and European, used hybrid mechanisms, such as transnational projects. Japanese firms also used informal coordination methods, which he considers highly valuable and underutilized in European companies.

In addition to the problems of integrating the activities of laboratories with other functions in the firm, such as marketing and production (which will be examined in subsequent chapters), another key MTI problem is ensuring efficient technology transfer between laboratories. A UK study conducted ten in-depth case studies of leading European chemical/pharmaceutical firms, of which six are among the top ten spenders of R & D in the industry worldwide (Senker *et al.* 1996). It examined technology transfer between laboratories and revealed that few companies had developed satisfactory methods. Each company organized regular meetings of senior research directors from its various laboratories. The meetings were used to discuss research programmes and reports, strategy, regulations, and alliances. Companies exchanged research reports (although these were sometimes in languages other than English). Short-term exchanges of staff between laboratories for a few weeks were quite

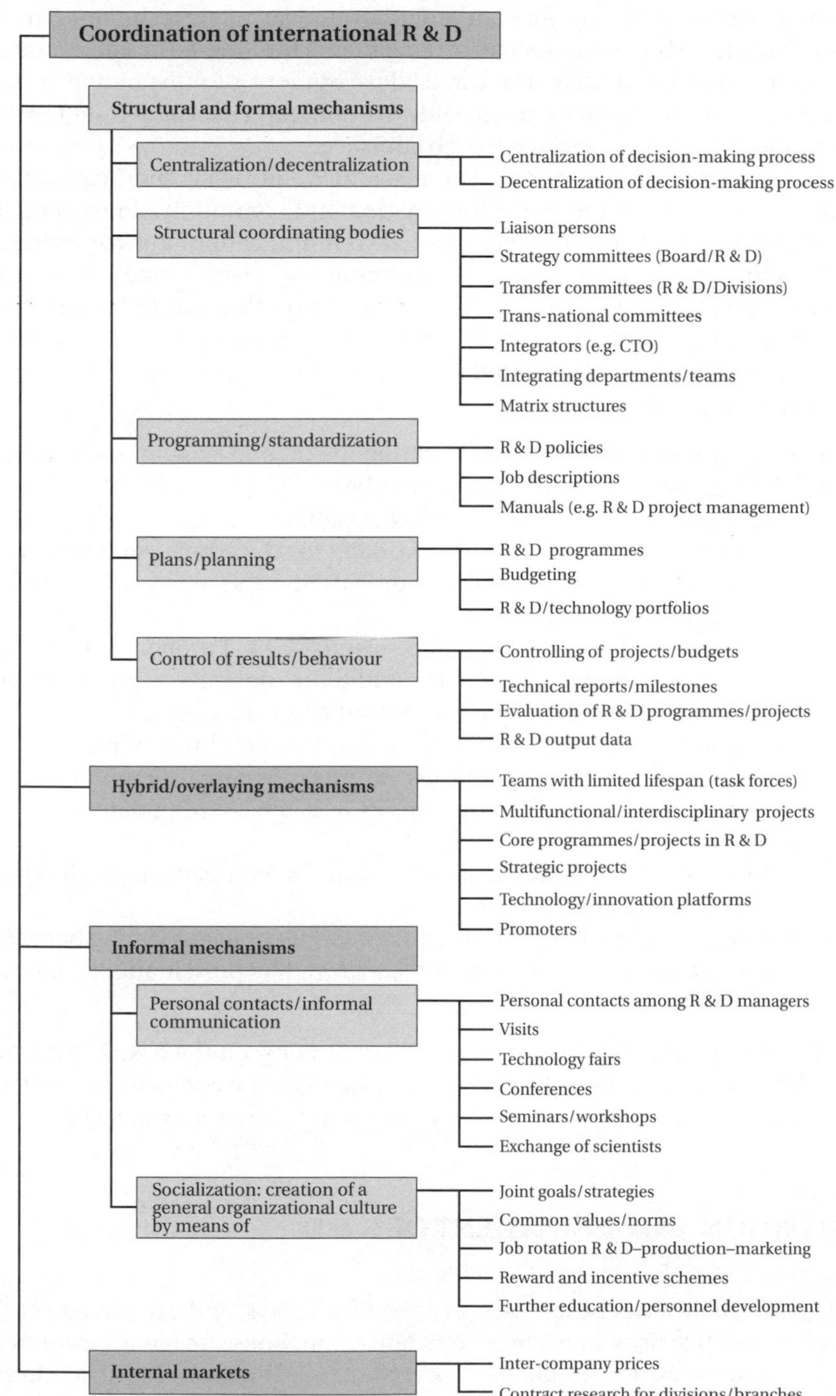

Fig. 3.9. Coordination of international R & D

common, with staff usually going to other labs to learn new techniques, or to train colleagues. Six companies had arrangements for long-term international placements, usually to assist the career development of outstanding young scientists, but these were infrequent. Only five companies seconded staff to the laboratories where they placed research contracts.

In common with other studies, Howells and Wood (1993) and Reger (1997) found little evidence of the use of information and communications technology (ICT) to facilitate effective international communications between laboratories, although e-mail and teleconferencing were used. The UK chemicals/pharmaceuticals study identified one firm that was the most effective at integrating its UK and USA laboratories. The company was trying to integrate its research between the different labs and their partners, using a number of strategies, including

- matrix organization (organized by discipline on one axis—biotechnology, cell biology—and disease area on the other);
- frequent meetings between senior R & D staff;
- major investment in electronic networking for the whole company, with twenty-four-hour computer communications and every US scientist linked to every European scientist;
- a company rule that every time a US scientist visits Europe, or vice versa, he or she has to spend two or three additional days getting to know the scientists with whom he or she is communicating electronically;
- construction of a research database (with access determined by seniority) containing research programmes annual review documents and summaries of all research programmes and quarterly updates of their progress;
- extensive use of scientific meetings within the company, using internal and external speakers;
- many short-term exchanges of scientists and technicians lasting between two to three weeks to several months, followed by presentations on what they have learned.

Despite these efforts, differing cultures still exist between the R & D laboratories. While these can constrain the flow of knowledge, the company also recognizes that cultural tensions can have creative results (Senker *et al.* 1996).

EVALUATION AND ASSESSMENT OF R & D

R & D performance can be assessed in a variety of ways, and can include evaluation of effective use of resources, efficiency, timeliness, revenues from new products/processes, and production cost reductions (Roberts 1995). Traditional investment evaluations cannot be used for R & D. Capital budgeting approaches like net present value and return on investment are difficult to calculate for R & D and can significantly understate the value of R & D investments. Discounted cash-flow analysis fails to include any long-term returns to R & D, and it is difficult to attribute a particular contribution to R & D investment. How much was a profitable new product the result of input from R & D, from sales and marketing, from manufacturing, or from business and collaborating partners? Some companies, like Hitachi, attempt to assign a profit contribution to each project—a notoriously difficult process. Companies like Philips consider a number of other methods, including

- determining the value of businesses that would not have been created without investment in R & D;
- measuring licensing and cross-licensing income from the company's patents (patents from Philips Research, the 3,000-strong central research function, earn enough in licensing income to pay for the research facility);
- considering option values (Carrubba 1993).

Some companies are evaluating R & D as opportunities, or options, that confer rights but not obligations to take some action in the future. The Executive Vice-President of Philips Electronics, for example, sees a number of advantages in using option pricing theory in R & D, because

> options (calls) limit downside risk. If for some reason it is decided not to make the follow-up investment needed to capitalize on a research project, only the research cost is lost, not the cost of follow-up investment. Actually, one might even recuperate some of the loss through patents. Furthermore, discovering—before your competitor does—that a certain technological option is a dead end, can be quite valuable. Whereas uncertainty (risk) decreases the Present Value of future cash flows, uncertainty (volatility) increases option values. (Carrubba 1993)

R & D option techniques emphasize investment values, which alter over time and across projects, as well as perceived investment cost and discovery value volatility (Newton *et al.* 1996). Research into the option values of R & D projects will provide an important future development. Some interesting findings have emerged already.

- R & D expenditures may be valuable even though no intrinsic value emerges, although this normally would not justify an R & D budget.
- Volatility of both investment cost and discovery value increases option value, since the actual investment may be deferred or cancelled.
- Lower or even negative correlation between cost and discovery value (especially in the case of highly volatile investment costs and discovery value) increases the option value of R & D, a consideration not always at the forefront of expenditure decisions (Newton *et al.* 1996).

Research managers need to be conscious of what the management literature has long understood as the problems of goal displacement as efforts shift towards ends that are most easily quantified (Hrebreniak 1978). In *The Reckoning*, an account of the havoc wreaked on Ford by Robert McNamara and his form of managerialism, which insisted on quantifying virtually everything, David Halberstam describes (in a manner no doubt resonant with many present-day researchers and engineers) how Ford engineers and R & D workers

> gradually [came] to doubt themselves. Repeatedly beaten on certain kinds of request for their plants over the years, they came to realize that some items would not go through, no matter how legitimate, and so they began to practise a form of self-censorship. They would think of something they needed, realize they could not get it through, and cut the request down so severely that the original purpose was sacrificed. (Halberstam 1987: 498)

McNamara's faith in systems analysis and statistics over any form of qualitative assessments was later, of course, to have even more devastating impact in the tragedy of the Vietnam War.

Whether the results of basic research can be quantified at all is open to question. In a review of US experience of studies on the benefits of basic research undertaken by, amongst others, the US Department of Defense, the National Science Foundation, and the Office of Technology Assessment, Steinmueller (1994: 60) concludes that, 'whether significant returns to public basic research investments in specific projects can be measured or predicted with significant accuracy to guide research investments remains an open empirical and conceptual question'.

There are numerous problems with measuring research outcomes, not least of which is the lengthy timescales involved in commercializing research. Furthermore, how do you assess the relative contribution of new, as compared to existing, know-how?

Aware of the problems of measuring the returns to research of all types, companies are careful to ensure that the most appropriate methods of research evaluation are used for different forms of research. Japanese firms such as Mitsubishi Electric, NEC, Sharp, Toray, and Hitachi evaluate their longer-term research according to its 'fundamentality' and the way it contributes to knowledge. Another method that can be used to assess R & D outcomes is an R & D effectiveness index (see Box 3.6).

CONCLUSIONS

R & D is a key source of competitiveness but, as this chapter has shown, it is difficult to organize and manage. This difficulty reflects the broad range of objectives of R & D, the different kinds of skills and personnel involved in it, the difficulties in measuring its outcomes, and the increasing challenge of globalization.

One of the major trends seen in the organization of R & D in the 1990s was an increasing move towards decentralization. The aim of this move was to make R & D better targeted to immediate business needs. The danger in this

Box 3.6. ***R & D effectiveness index***

$$EI = \frac{\% \text{ New Product Revenue} \times (\text{Net Profit } \% + \text{R \& D } \%)}{\text{R \& D } \%}$$

Ratio of increased profits from new products divided by in product development

EI > 1.0, return greater than investment

e.g. 9% net profit, 6% investment in R & D: 40% revenue from new products → EI = 1.0; 20% revenue from new products → EI = 0.5

- Like all metrics the objective is relative comparisons, not absolute performance (and can be used across divisions and industries, and for benchmarking)
- Definition of new products (suggested to be those in the first half of product life cycle)
- EI is driven by time-to-market and product success

Source: McGrath and Romeri (1994).

move is that it underemphasizes the importance of longer-term, more speculative R & D that provide the opportunity to create new markets, and give the customers what they did not know they wanted. Historically, there has been a cycle of centralization and decentralization and this may well be repeated, as the recognition of the benefits of centralized research returns. The long-term competitiveness of R & D requires a balance of near- and long-term R & D, with a proportion of the R & D project portfolio being in the high-risk, high-return category. As it is very difficult to assess the returns to R & D, a variety of new techniques are beginning to be used to evaluate the benefits. This includes the relatively new field of assessing R & D as 'options'.

R & D management requires both team management and individual creativity, and the issues of recruiting, rewarding, and organizing teams and individuals are central. R & D workers tend to be highly educated and skilled, and the most effective rewards tend to be more complicated than simple remuneration. High levels of task discretion and autonomy are common, but there is a need for formal organizational decision-making processes. These project management systems take a variety of forms, including those related to specific individuals at particular stages of the project's development. Integrated teams with a common system or product focus were shown to be effective in the computer and auto industries. The tendency towards the increased internationalization of R & D has many consequences for its management. The organization of international R & D, its reporting structures, funding, and management styles, depends ultimately on the strategy of the firm, an issue to which we return in Chapter 6.

CHAPTER

4

The Management of New Product Development

One of the primary aims of companies' investment in research and development (R & D) is to produce new and improved products. As we saw in the previous chapter, the results from R & D investment can be intangible. They may, for example, increase the understanding of scientific principles underpinning a company's products, and can provide options for the future through the creation of intellectual property rights (IPR). They may improve the company's ability to manufacture new products. The major aim of corporate R & D, however, is to combine technological and market opportunities to produce differentiated goods and information (new and improved products and services) that convey competitive advantages. These products can be sold or licensed for royalty income.

WHAT IS A NEW PRODUCT?

Products can be new to the world and new to the firm. A product new to the world would be, for example, Hewlett-Packard's (HP) Laser Jet Printer: it was the first ever produced. When IBM introduced its laser printer, it was a new product line for the firm, but followed HP's product onto the market. New products can add to existing product lines, such as when HP introduced its Laser Jet IIP, or when Boeing introduced the 747:400, and they can be improvements to existing products, such as Microsoft's various versions of Windows. Products can also be repositioned into new markets. An example would be the use of aspirin as an anti-stroke agent as well as a pain-reliever. These different categories are shown in Fig. 4.1.

The potential benefits of new product development

According to Clark and Wheelwright (1993: 83–4), the potential benefits of effective new product development (NPD) are threefold.

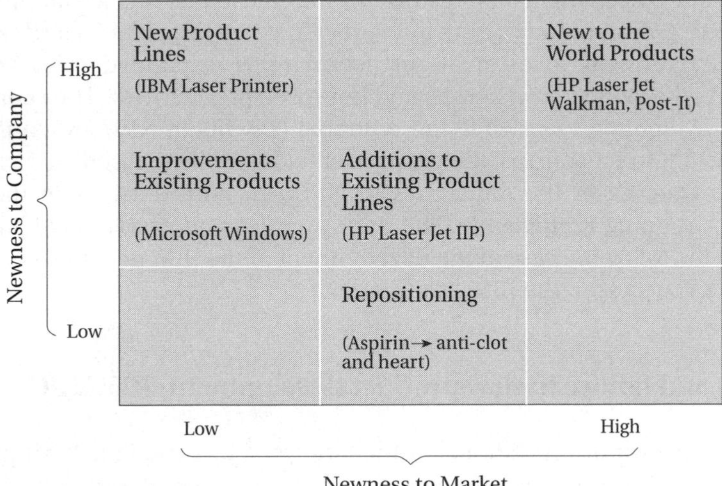

Fig. 4.1. Categories of a new product
Source: Cooper (1993).

First there is the question of *market position*:

Ideally a new product can set industry standards, standards that become a barrier to competitors or open up new markets, such as the Sony Walkman or Polaroid camera. Superior products and processes are a means to get the jump on competitors, build on existing advantages by creating stronger competitive barriers, establish a leadership image that translates into market dominant designs, extend existing product offerings, and increase market share.

However, new products often require new marketing channels which can present challenges to many firms.

Secondly, there are benefits of *resource utilization*, which can include

capitalizing on prior R & D investments (applying lab discoveries), improving the returns on existing assets (such as the sales force, factories and field service network), applying new technologies for both products and manufacturing processes, and eliminating or overcoming past weaknesses that prevented other products or processes from reaching their full potential. The potential leverage on a variety of resources can be substantial.

Finally,

The excitement, image and growth associated with product and process development efforts capture the commitment, innovation and creativity of the entire organization. This success, in turn, enhances the firm's ability to recruit the best people, improve their integration, and accelerate the pace of change. Furthermore, development projects themselves often are the vehicle by which new approaches and new thinking are adopted and take on institutional reality.

Clark and Wheelwright (1993) contend that these benefits are additionally anticipated to 'drop to the bottom line, providing rich financial rewards such as improved return on investment, higher margins, expanded sales volume, increased value added, lower costs, and improved productivity'. They continue, however, by arguing that, while the potential benefits of NPD are enormous, unfortunately, in most firms, the promise is seldom fully realized.

This is because of the difficulties in NPD—in both developing the right product and doing it efficiently. The opportunities and problems of NPD are shown in the following case of the development of the IBM personal computer (PC) and its successor, the IBM PC Junior.

Success and failure in new product development: IBM's PC

The development of the IBM PC in the late 1970s was the fourth attempt by the company to produce a microcomputer; all previous attempts had failed. The PC was developed in a small company operation at Boca Raton, Florida, with a small development team of eleven. It used a highly entrepreneurial approach to development. The team was put under tight time constraints and given a year to complete the project. The development involved detailed market research, with the development team visiting many computer gurus and marketing consultants. It also operated in very different ways from the traditional 'IBM way'. Contrary to the company's usual practice of designing and making everything in-house, in this case important expertise and component elements were bought in. For example, the team involved a small software developer to design software, and its central processing unit (CPU) was designed elsewhere. Close attention was paid to the product's market-entry strategy by carefully establishing the price/performance balance and target- market niche (the cost of the computer was to be $3,270).

The IBM PC was a huge success as a new product. Introduced in 1981, it sold 538,00 in 1983 and 1,375,000 in 1984, capturing 40 per cent of the PC market. As soon as production capacity caught up with the demand for the product, IBM cut the price of its PC by 23 per cent. This move was viable as high production volumes produced learning curve and scale economies, and thereby enabled high margins. The price cut was aimed at increasing market share and deterring imitators with smaller-scale production capacity.

On the other hand, the IBM PC Junior (developed subsequently) was a failure as a new product. It was developed at what had grown into a large operation at Boca Raton with over 10,000 employees. The entrepreneurial ethic was replaced by a bureaucratic approach to development, with a complex structure of internal product design and marketing teams. There were few market linkages as a basis for establishing design specifications, and little consultation with outside experts. The Junior was designed to avoid competing with the successful PC and not on the basis of identified user requirements. Its 'toylike' keyboard made it highly unsuitable for word processing (which was to be the main office application) and too costly for the home market (it was priced at $1,000 as opposed to $200 for the top-selling Commodore 64). It also had limited memory, which rendered it unsuitable to compete at the upper end of the market. As a result, the Junior was 'neither fish nor fowl' and very few machines were sold. Subsequently, IBM introduced an optional typewriter keyboard for the Junior with a conventional layout, expanded its memory from 128,000 characters to 256,000 characters, and cut the price of existing models. None of these efforts saved the product and eventually it was withdrawn from the market.

As new products, the PC was a success and the Junior a failure. The market orientation, entrepreneurialism, and preparedness to integrate external inputs that characterized the PC's development was not continued into the development of the Junior. But what was the impact of the development of the PC on IBM as a whole? The software supplier used by the PC development team was Microsoft; the chip provider, Intel. These two companies have subsequently defined the development of the personal computer market and have grown to be larger than IBM itself. IBM developed a good product by using other companies' proprietary technology. The technology, however, was not protectable by IBM and sales of the IBM PC collapsed with the arrival of PC clones from companies using the same technology such as Compaq and Dell. Ultimately, IBM failed because it believed it could do a better job than its initial suppliers and therefore missed the opportunity to acquire or contractually limit Microsoft and Intel. By 1990 IBM had less than 10 per cent of the personal computer market. The IBM PC was a success, in so far as it assisted IBM to achieve its strategy of moving out of mainframes and using outsourced components. Nevertheless, the strategic advantage was lost because it failed to control the intellectual property needed to do so. These strategic issues of NPD will be examined in later chapters.[1]

WHAT MAKES A FIRM INNOVATIVE IN ITS NEW PRODUCT DEVELOPMENT?

The NPD process requires firms to be both creative and controlled, to share information but also to protect it as a major source of competitiveness, and to respond to existing market demands as well as produce differentiated products using novel technological advances. The complexity of the NPD process and the difficulties firms have in dealing with it have invited a huge literature on the subject. In the automobile industry alone, for example, major studies such as

[1] This discussion is based on various sources, most particularly discussions with Paul Gardiner, and also Cringely (1996).

Womack *et al.* (1990) and Clark and Fujimoto (1991) have produced enormous detail on the process of producing new cars. Some of these studies have focused considerable attention on the Japanese method of NPD as an efficient means of generating and producing new products. Micro-level studies have shown that Japanese firms spend much less than Western counterparts on R & D, yet manage to register more patents, and econometric modelling has shown that applied R & D has yielded a higher rate of return in Japan than in the United States (Mansfield 1988*b*).

As the model of Brown and Eisenhardt (1995) shows (Fig. 4.2), it is important to distinguish between process performance (how efficient is a firm at developing new products?) and product effectiveness (is the firm producing the right products?). The latter is essentially a strategic issue and will be examined in Chapter 6. Here the focus will be on those factors that improve process performance, including internal organizational integration, human resource management issues, and tools and techniques of project management.

Internal organizational integration for new product development

The close integration of the management and organization of NPD is critical to its success (Rothwell 1992). The best international benchmarks of NPD performance include Japanese auto and electronics companies, such as Toyota (whose practices are being emulated around the world) and Sony, small US high-tech companies typified by those in Silicon Valley, and larger corporations, such as 3M, that structure themselves to encourage entrepreneurial activities.

There are striking similarities in the NPD process in large Japanese firms and the model found in smaller Silicon Valley high-tech firms (Nonaka and Kenney 1991). These include the way NPD is focused in a comparatively small group

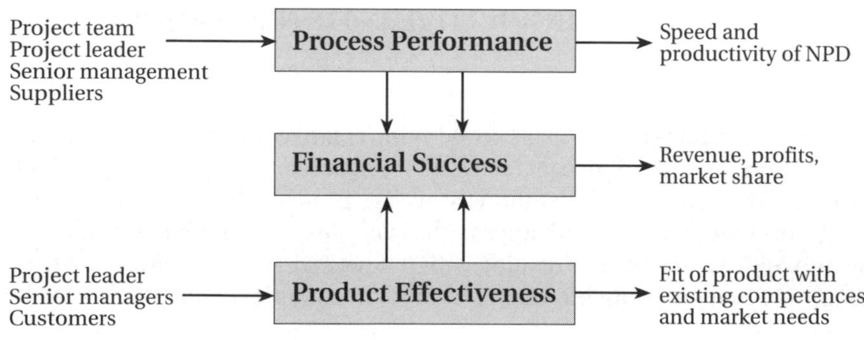

Fig. 4.2. A new model of product performance
Source: Brown and Eisenhardt (1995).

that is intensely interactive and committed. This organizational model is particularly conducive to information sharing and good communications. This form of organization includes the functional integration of teams 'concurrently engineering' new products. Imai *et al.* (1988) use the analogy of a relay race to describe Western NPD, with each function (marketing, R & D, manufacturing) becoming involved sequentially. In contrast, they use the analogy of rugby to describe Japanese NPD, where the different functions with different capabilities simultaneously attempt to reach the same goal. Concurrent engineering is now commonly practised in US firms, with companies like Hewlett-Packard substantially reducing the development time of new products through its use. The development time for printers, for example, was reduced from fifty-four to twenty-two months (Meyer 1993; Zangwill 1993).

Effective managerial and organizational integration includes the minimization of reporting layers and the encouragement of multifunctional and cross-divisional links in order to improve links between R & D, manufacturing, and marketing (Whiston 1991). While there are many advantages in using the flat, organic organizational structures seen in Japan, these structures are very different from and challenging to traditionally organized large firms in other countries (Tushman and Anderson 1986; Henderson and Clark 1990). However, the success of Japanese NPD processes had led to emulation in other parts of the world, and many US firms (like Du Pont and Motorola) are now using the principles of concurrent engineering. Boeing developed a new philosophy called 'Working Together' so as to gain the benefits of organizational integration in the development of the 777.

3M uses different organizational structures for the development of the different types of product that achieve its vision statement: 'To be the most innovative enterprise in the world, and be the preferred supplier'. Its view of NPD development follows that of Prahalad and Hamel (1994), who distinguish between firms that develop products that customers do not want, those that develop new products customers do want, and those that develop products customers do not know they want. To develop products customers want, 3M uses an organizational structure with a high level of discipline, accurate schedules, and a team that is usually large, with a low-risk/low-failure profile. To develop new technologies that may develop products customers do not know they want, 3M uses creative individuals, with little management involvement, operating with a high-risk/high-failure profile (Nicolson 1998). This is encouraged by the '15 per cent rule', which allows all 3M staff to spend 15 per cent of their time working on their own projects.

Successful NPD almost always requires close integration with the marketing function. Successfully innovative firms have a strong market orientation, with an emphasis on satisfying user needs and creating product advantages such as unique benefits, high quality, attractive price, or innovative features. (Generally, new products succeed when emphasis is placed on economic benefits rather than technological novelty.) Good-quality information from the market is essential to understanding these user needs, and many innovative firms actively involve customers in NPD. It is important to understand buyer behaviour and customers' purchase decision processes in, for example, the balance of price/quality/delivery time sensitivities. Some companies have a targeted sales strategy of identifying early and high-profile adopters, who can then be cited in sales literature. The level of attention to customers is seen in one of 3M's key corporate values of

satisfying our customers with superior quality and value [by]

- providing the highest quality products and services consistent with our customers' requirements and preferences
- making every aspect of every transaction a satisfying experience for our customers
- finding innovative ways to make life easier and better for our customers.

A brief note on the dangers of listening exclusively to marketing departments

Every company has its 'war stories' of corporate events, and individuals have their own selective memory of how things occurred. With these provisos in mind, this is an account of 3M's development of the Post-It note, related by one of the key personnel involved.

The key innovation in the Post-It note is the adhesive—the clever, non-sticky glue. Initially, the development team did not know how best to use the adhesive and they covered noticeboards with it to which people could attach pieces of paper and messages. The idea of putting the glue on a small note eventually came up and was suggested to the marketing department. After conducting market research, the department said there was no demand for the product. The development team persisted and the marketing department twice more rejected the idea. Someone in the development team had the idea of sending the prototype notes to all the secretaries of 3M's top managers. Shortly afterwards, secretaries began to telephone to ask for more stock. The development team redirected the enquiries to the marketing department, and the marketing department quickly learnt that there was a demand amongst some of the most influential people in the company. The product was thereafter launched.

Another example of ignoring the market department is provided by Sony, which produced over 300 different versions of the Walkman. These versions, however, were not developed in consultation with the marketing department, but rather were produced and examined to see which models sold (Hayes and Pisano 1994). Such cases are, however, exceptional, and more commonly strong marketing input is needed in the NPD process (see Chapter 8).

The importance of coordination in new product development

When customers are involved in the development of a new product they can valuably place demanding conditions on the development team. Tough, demanding customers usually produce good designs and products. Rolls Royce, for example, attributed the success of one of its aeroengines to the demanding requirements of Boeing (Gardiner and Rothwell 1985). Links with customers continue after the products have been purchased. Efficient after-sales service not only keeps customers happy but also allows sales and service staff to become a major source of information on future product improvements.

Integration with the production and operations, or manufacturing, domain is similarly important. In innovative firms this domain provides high-quality and well-planned production sequences with great attention to quality control, consistent meeting of delivery target and flexible production sequences to respond to changing customer requirements (see Chapter 5). Innovative firms also have close proactive links between production and NPD processes. As Pisano and Wheelwright (1995) put it, 'it is not only possible to excel at simultaneously developing new products and new manufacturing processes but also necessary'. New products have to be designed for production—that is, they have to be capable of being made as efficiently and cheaply as possible. A story circulates about a famous British chemical company that built a pilot plant for a new polymer. As a result of a combination of unexpected events the polymer solidified, seizing up the plant. The production engineers contacted the scientist who had developed the polymer and learnt that he too had experienced the problem. Mightily relieved, the engineers asked what he had done to solve the problem. They probably did not want to hear his answer: 'I smashed the test tube.'

Compared with the situation in US or European firms, in a Japanese firm the relationship of the R & D division with the production, marketing, and other divisions is extremely close. The (advanced) engineering departments of Japanese firms play a significant role in stimulating NPD (Aoki and Rosenberg 1987: 13):

- The engineering department is physically located on the manufacturing site, and engineers assigned there normally have a good command of practical knowledge concerning the manufacturing process.
- The primary responsibility of the engineering department is to develop and apply the engineering know-how of the manufacturing division to related uses.
- The relative importance of the engineering department of the manufacturing division, located at the manufacturing site, strengthens the reliance upon the localized use of on-site knowledge.
- This relationship also facilitates a close interchange of information between those responsible for product design and those responsible for the manufacturing technology.
- Additionally, it facilitates an easy communication among professional specialist groups who have separate but closely connected responsibilities.

NPD should also involve manufacturing and finance divisions of a company. There are advantages in extending the close liaison between NPD efforts and manufacturing to links with shop-floor workers. After Ford had designed the Taurus, 120 assembly workers were brought together to build 200 prototypes of the car. They suggested no less than 700 improvements in the design to assist manufacture (*Business Week*, 24 July 1995). Similarly, integrated NPD activities have to include financial managers. Finance teams can offer valuable advice on R & D and operations, and there are advantages in involving them in the planning of research-intensive periods of projects. Basically, successful NPD requires coherence and coordination across all levels of a company's activities. Clark and Fujimoto (1991: 7) argue that 'what seems to set apart the outstanding companies in product development is the overall pattern of consistency in their total development system, including organizational structure, technical skills, problem solving processes, culture, and strategy. This consistency and coherence lie not only in the broad principles and architecture of the system, but also in its working level details.'

Bowinder and Miyake (1993) refer to the range of mechanisms used by Japanese firms to ensure this functional and organizational integration, including in-house, on-the-job training and job rotation. Kenney and Florida (1994) found that the transfer of employees between R & D and manufacturing, and between basic and applied research, provided the most important mechanism for connecting these activities.

Internal integration is facilitated by, and encourages, the receptivity of staff to new information. The expectations of a typical Japanese R & D worker is that he or she will work with people operating in the manufacturing and marketing functions, and people from these functions may become colleagues as others move into the R & D function or R & D workers move into other functions. As such, there is likely to be a high level of preparedness to work with, and relate to, people outside their department and functional speciality. This preparedness also applies, of course, to small firms generally, where functional divisions are rare and staff have to be multi-skilled. Job rotation will be examined in more detail in a following section.

A brief note on product families and platforms

Research into NPD has shown the value in having 'families' of related products, which build on core technologies and have the flexibility to adapt to changing customer requirements. Various terms are used to describe this practice, ranging from 'robust designs' (Rothwell and Gardiner 1988) to 'product platforms' (Meyer and Lehnerd 1997). A classic case of a product family is Rolls Royce's RB211 aeroengine. The original engine was designed so that it could be 'stretched' into different product configurations. Engines could be offered with improved performance (increasing thrust), re-rated (taking advantage of technical advances), or de-rated (reduced thrust for improved fuel economy) (Rothwell and Gardiner 1988).

The hugely successful Sony Walkman also provides an example of an effective approach to the management of a product family—management practices that have led to continuing market leadership since its introduction.

Sony's strategy employed a judicious mix of design projects, ranging from large team efforts that produced major new model 'platforms' to minor tweaking of existing designs. Throughout Sony followed a disciplined and creative approach to focus its sub-families on clear design goals and target models to distinct market segments. Sony supported its design efforts with continuous innovation in features and capabilities, as well as key investments in flexible manufacturing. Taken together, these activities allowed Sony to maintain both technological and market leadership. (Sanderson and Uzumeri, 1995)

Based on their study of the Walkman, Sanderson and Uzemeri (1995) suggest that there are four tactics of product planning that may be applied to other manufacturers of incrementally improving products.

- *Sony pursued a variety-intensive product strategy, developing a large number of models and often pre-empting the competition.*
- *It decentralized decision-making about new products. Design teams were cross-functional and changes were periodically led by industrial designers and marketers depending on the nature of the problem being faced and the quality of information they had attained (such as in the key US leisure market).*
- *The company used its industrial designers judiciously, so that they could exploit their creative talents for introducing incremental changes.*
- *Sony minimized design costs by building all models around key modules and platforms.*

The use of common and related components also allowed the company's investment in flexible manufacturing equipment to be amortized over a longer period.

Human resource management factors in new product development

The truism that success in business is people centred applies to NPD. As well as the creative scientists and engineers discussed in the previous chapter, NPD needs good managers. While formal management techniques can enhance the performance of average managers, they are no substitute for high-quality managers who are good at managing people and change. There is a need for dynamic, open-minded managers who exercise 'subtle control' by formulating and communicating a vision of a distinctive, coherent product concept and delegating sufficient autonomy to encourage motivation and creativity (3M calls the latter a 'permissive attitude of managers'). Success in NPD often depends on a negotiated balancing act between strategic product vision and tactical problem solving, and NPD managers have to be able to do both.

NPD managers need to be good at a variety of activities, including:

- managing teams, some of which can be very large—the Ford Taurus team, for example, had a development team of 700;
- establishing clear policies and guidance about technical matters and coordinating technical/design issues—Iansiti (1993), for example, discusses the importance of frequent and rapid changes in design to deal with turbulent environments and the need for systems knowledge that describes the interactions between the broad architecture of products and specific design details;
- boundary spanning/communications with other domains, particularly with manufacturing and marketing (Fairtlough 1994*a*);
- lobbying/being an 'ambassador';
- gaining, protecting, and managing resources and IPR;
- creating an 'impression';
- along with top managers and product champions, creating a 'vision' and clear set of goals for the new product.

NPD managers also need to be confident of top management commitment to, and visible support for, innovation. Studies have shown the importance of enthusiastic and powerful project or product 'champions' or sponsors (with significant decision-making responsibility, authority, and high hierarchical level) (Rothwell *et al.* 1974). So, for example, the team that designed the Ford Taurus wanted to make the car's body from a single piece of steel, rather than welding two pieces together, thereby reducing welding and air noise, and producing greater rigidity and better fitting doors. The manufacturing function balked at the cost of a new 7,000 ton press at a cost of $90 million. The Taurus team eventually persuaded the company's chairman to make the investment, despite his reservations about the cost. He believed strongly in the new car and was a driving force behind it (*Business Week*, 24 July 1995).

A significant human resource management (HRM) challenge is, therefore, to attract, reward, and retain these talented managers and, through a commitment to human resource development and training, to improve the management skills of NPD personnel. There is a variety of incentives available to encourage NPD managers, ranging from financial rewards, including salary increases and stock options (favoured amongst small high-tech firms such as those in Silicon Valley), to peer recognition (seen in the case of 3M and its prestigious Carlton Society Awards). Oxford Instruments, a scientific instruments company in the United Kingdom, encourages managers who develop new products to spin them off as related companies with the manager as CEO.

Job rotation is an underemphasized, but hugely important, HRM practice in NPD, and it is worth examining its role at some length. In Japanese firms job rotation is crucial to information flow, organizational coherence, and hence the NPD process. There is a four-stage job rotation for Japanese R & D personnel (Wakasugi 1992):

- The first stage of the career path is an educational period, which begins when the recruit joins the firm and lasts until he or she reaches late 20s. The recruit provides support for R & D activities during this stage.
- In the next stage, the early 30s, the employee becomes a fully-fledged engineer.
- In the third stage, between the age of 35 to around 40, the employee becomes a manager at the front line in the research division. (This is usually a section chief, project leader, or comparable level.)
- Finally, the employee leaves the forefront of R & D activities and joins general management.

During the third and fourth stages, employees are often assigned to manufacturing and marketing divisions, and they experience what Wakasugi calls a 'spiral promotion process', allowing them to learn about the corporate decision-making process or about product planning, production, and marketing. Beginning in the 1990s, some Japanese firms have also introduced elements of dual career paths, so that senior scientists can remain in the laboratory and be promoted and rewarded to similar levels as managers (a practice common in the West).

Aoki (1988: 238) describes the way this rotation of staff assists innovation as follows.

> If a potentially promising development project proposed by the engineering department requires more basic scientific knowledge than the department possesses, the project may be commissioned to the central research laboratory. When this happens, one or more young engineers from the engineering department are dispatched to the central research laboratory to participate in the project team. After the team has solved the basic problems, the project is handed back at a relatively early stage to the engineering department of the commissioning manufacturing division for detailed design and testing. In the process, the dispatched engineers are transferred back to the original department. They implement research results, by converting them into a manufacturing process.

A survey of heads of R & D divisions and middle management in Japanese firms shows that very few have worked exclusively in R & D (Wakasugi 1992). Only 10 per cent of R & D recruits in NEC ultimately remain in research. The rest move to other divisions, often to implement some of their own concepts or developments. Engineers in Epson spend 70 per cent of their time in design and 30 per cent in marketing (Subramanian 1990). The job-rotation process not only brings new information and technology, but also, as it is a frequent occurrence, improves the receptivity of staff to new ideas and technological novelty. It improves the opportunities for, and likelihood of, learning.

Project management techniques in new product development

While rigorous planning and control are essential in stable, complex, and relatively mature sectors (such as most elements of autos and computers), it is perhaps less important in uncertain and fast-moving sectors where learning and experimentation are more suitable. However, all NPD projects require regular appraisal and the commitment of resources to upfront screening of new projects and pre-development activities.

One well-known system of appraisal is the stage-gate system developed by Cooper (1993), and used by companies such as Exxon Chemicals, Corning, Ericsson, Kodak, Carlsberg Breweries, and Procter & Gamble. Essentially, they systematize decision-making at various stages of a new product's development. This involves:

- stop/go decisions at each stage;
- increased resource commitment at each stage;
- evaluation by appropriate staff (with increasing levels of seniority at later stages);
- non-partisan, fair decision-making.

In Cooper's generic stage-gate process there are five key stages:

- *preliminary investigation*: a quick examination and scoping of the project;
- *detailed investigation*: a more detailed consideration leading to a business case, including project definition, justification, and plan;
- *development*: the actual design and development of the new product;
- *testing and validation*: tests and trials in the marketplace, laboratory, and plant to verify and validate the new product, and its marketing and production;
- *full production and market launch*: commercialization, the beginning of full production, marketing, and selling.

He adds two other stages, idea generation and strategy formulation, which are outside the actual process of NPD.

Gates are predefined with established mandatory and desirable criteria to be met. They bring a valuable discipline to the process of NPD, not only in the sense of providing a sound basis for resourcing competing projects, but also in ensuring that marketing input is continually integrated into a process that sometimes can be driven by technological possibilities rather than customer focus.

The developer of the stage-gate system, Bob Cooper, argues that around 40 per cent of firms introducing it experience difficulties, although benefits have been achieved. The problems lie in implementation—exerting the level of discipline required. He also provides some insights into the future development of stage-gate systems designed to improve their efficiency. These encompass the following.

- *Fluidity*. Increasingly, fluid systems will emerge whereby a project can be in two stages at once. So, key activities of Stage 3 might begin before Stage 2 is completed, reducing lead times and accelerating the entire process.
- *Flexibility*. Stages will be combined, gates collapsed, and activities/

deliverables omitted, especially in the case of lower risk projects. As Cooper puts it, 'Stage-gate must be a roadmap, not a straightjacket.
- *Fuzzy gates.* In traditional stage-gate processes, gates are either open or closed. With fuzzy gates they can assume various states in-between. That is, a project can proceed into the next stage with partial information (a 'conditional go'), depending on certain events occurring or information becoming available.
- *Focus.* Cooper's research finds many of the problems besetting product development are the result of poor project prioritization, too many projects, and a lack of focus. The answer, he believes, lies in effective portfolio management, which examines the range of projects being conducted and considers them in respect of maximizing value, achieving the right balance, and linking them with business strategy (Cooper 1994; Cooper and Kleinschmidt 1996; *Innovation Management Network, URL:www.aom.pace.edu/tim*—16 Dec. 1996). The management issues in linking stage-gate systems with portfolio management and business strategy are difficult, and remain a continuing problem for many firms.

A variety of mechanisms can be used to focus staff towards NPD. Perhaps the most dramatic is 3M's strategic aim of gaining 30 per cent of sales from products developed in the past four years. Some Japanese and Korean firms construct crises to focus the attention of the staff. So, for example, Canon originally had a simple strategic aim of 'being an excellent company in Japan'. (These slogans act as important focusing devices in Japan.) When all achievements and measurements indicated that it had achieved this goal, the goalposts were moved to 'being an excellent company in the world', which at the time it was not. Rather than complacently accepting that they had achieved their aims, Canon staff were challenged into setting higher objectives. In Korea, constructed crises are a common focusing device for firms (Kim 1997).

A brief note on bootlegging

Bootlegging involves highly motivated individuals working on innovative projects without official sanction and funding from management. It is seen as a valuable means of producing good ideas, unconstrained by formal organizational and project requirements. Augsdorfer's (1994) study of twenty-four British new-technology based companies found nineteen where acknowledged bootlegging occurred, and on average between 5 and 10 per cent of researchers undertook it. In one firm 50 per cent of researchers bootlegged. Pearson (1997) suggests that bootlegging can arise for a variety of reasons, including to satisfy curiosity, to attack technical problems that led to project termination, or to assess markets that the organization does not believe are viable or is not prepared to consider. He believes that bootlegging is such an important activity for innovation that management should play a more formal role in its conduct. Pearson also suggests that, whilst ICT enables more bootlegging outside work, by staff accessing company information remotely, budget constraints and greater formality in research management (such as total quality management (TQM) practices (see Chapter 5)) may restrict this valuable activity.

Failure in new product development

New products can fail in many ways. For example in the way they

- do not meet users' needs;
- are not sufficiently differentiated from the products of competitors;
- do not meet technical specifications;
- are too highly priced for perceived value;
- do not comply with regulatory requirements;
- compete with the company's other products;
- lack strategic alignment with the company's product portfolio.

Most new products fail in the marketplace. 3M, for example, claims to fail in over 50 per cent of its formal new product programmes. This high level of failure is shown in Fig. 4.3.

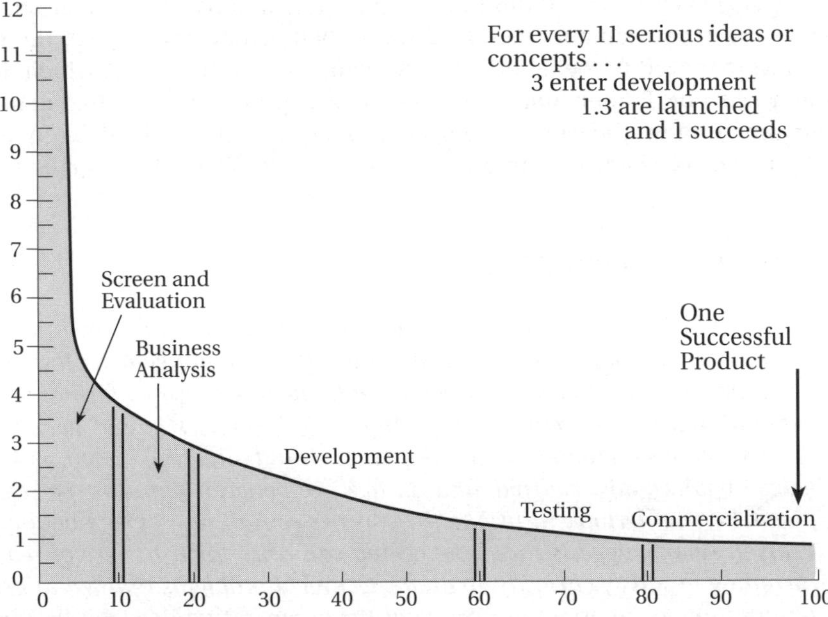

Fig. 4.3. Attrition rate of new product projects
Source: Booz *et al.* (1982), cited in Cooper (1993).

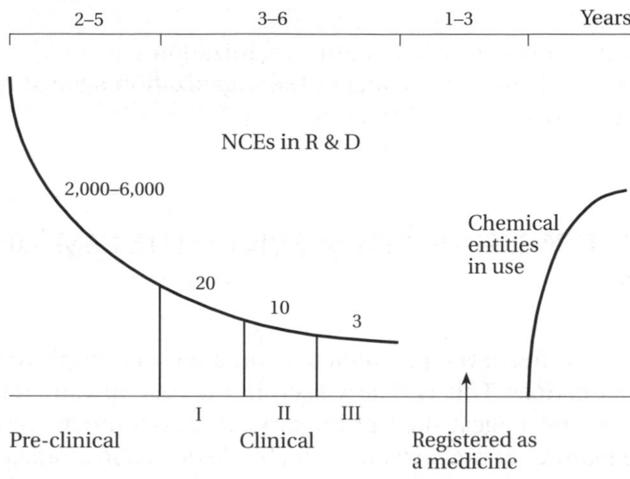

Fig. 4.4. New chemical entities in R & D

In the chemical industry, the attrition rate in the development of new chemical entities is very high. At any one time, between 2,000 and 6,000 new chemical entitities are being examined in pre-clinical trials. This number reduces to twenty when clinical trials (experiments with humans) begin. Only three will emerge at the final stages of clinical trials, and these will have to win a place in the market (Fig. 4.4).

NPD is a risky activity. The types of firms introducing new products face/include:

- *market* risks—uncertainty about demand;
- *competitive* risks—what will competitors do?
- *technological* risks—will the product work?
- *organizational* risks—what organizational changes are needed?
- *production* risks—can the product be made?
- *financial* risks—large up-front investments are required for uncertain pay-offs (Garvin 1992).

As we have seen, however, new products are important for firms and the greater risk for the future of a company may lie in not developing them. Furthermore, even failure produces learning and may be valuable. 3M sees its failures as a learning experience. This is a view shared by Maidique and Zirger (1985), who argue that past product failures provide the basis for

- new approaches to marketing;
- new product concepts;
- new technological alternatives;

- new information about customers that may lead to major product reorientation;
- indentification of weak links in the organization;
- inoculation of strong elements of the organization against replication of failure patterns.

THE DEVELOPMENT OF NEW SERVICES: THE CASE OF CITIBANK

Good practices in the development of new products also apply to the development of new services. This is clearly seen in the case of Citibank, one of the world's oldest and largest banks.[2] Its services development practices show many of the features of organizational, technological, and strategic integration discussed in respect of new products.

Banking is an industry where many large players offer similar products (deposits and loans). A major way of distinguishing between brands or institutions is through the perceptions of customer service and the technology that facilitates it. Citibank's business strategy calls for the use of technology-based products and services to take advantage of the changes occurring in the financial services and technology marketplaces, most notably the emergence of virtual banking and e-commerce. To ensure that its technology strategy and business strategy are integrated, it has established a Global Technology Group, whose role includes leading technology projects and building a 'technology acceptance culture' in the firm.

Citibank has succeeded in building a reputation for technological innovativeness. It employs over 8,000 technologists and took the lead, for example, in the development of automatic teller machines (ATMs). The Citibank Development Center is recognized in the industry as being a leader in the research, design, and development of advanced computer systems and products, and transaction-processing technology for the delivery of financial services. Unusually for a service company, it registers patents. The Citibank Development Center has registered twenty patents and in 1998 had another twenty-two applications pending.

One of the reasons for the technological focus of the company is the complexity of its global operations. Its Card Products Division, for example, handles 14 million line transactions daily, and its technical staff maintains over 20,000 program modules. The need to coordinate the enormous scale of these activities led it, in 1992, to be the first major US corporation to consolidate its global telecommunications networks into a single organization, the Citibank Global Technology Infrastructure. This enabled the delivery of

[2] This discussion has been based on information provided by Eugene Kwek.

products, services, and information to internal businesses and customers worldwide. In the late 1990s Citibank was building a common systems architecture, process, and language to support its global operations. This involved consolidating hundreds of mainframe data centres, partnering with major technology vendors and suppliers and implementing new, global standards for user computing. Technology also assists with security—Citibank has an initiative known as Security Architecture, which is a comprehensive information security program for the next generation of computer internet banking. Other projects include new touch-screen ATMs, cash acceptance machines, as well as quick-check deposit machines that give a receipt after deposits are made.

Citibank uses cross-functional teams in the development of new services. Staff from different sections and departments combine to maximize the transfer of information and technical skills. Staff are encouraged to join special task forces around new projects to lend their skill and expertise. End-users are also involved in the development of software and hardware integration, and suppliers are involved early in the development process to ensure the proper configuration of equipment.

Since the early 1990s, Citibank's software development efforts have sought to improve quality, reduce development time and cost, and establish processes for continuous improvement. To do this, the company uses process design and object-oriented programming in the pursuit of zero-defect reusable code. It uses Carnegie Mellon Software Engineering Institute's CMM. The CMM provides software organizations with guidance on how to gain control of their processes for developing and maintaining software. It was designed to guide organizations in selecting improvement strategies by determining current process maturity and identifying the issues most critical to software quality and process improvements.

A brief note on the importance of flexibility in design

Flexibility of product design is a key element underlying successful innovation (Iansiti 1993). Flexible organizational structures are equally important in enabling firms to react to changes in markets, new opportunities in production technologies, and new possibilities derived from developments in science and technology. Also important is the flexibility of the technology used in the design process. Thomke (1997) compared the productivity of two design technologies used in designing integrated circuits: the highly flexible electrically programmable logic devices (EPLDs) and the less flexible application-specific integrated circuits (ASICs). When they were used in the design of similar products, Thomke found that use of EPLDs outperformed ASICs by a factor of 2.2 (measured in person months). This he attributed to the more flexible design technology, allowing a greater degree of risk in design (a greater preparedness to remain open to design changes, rather than operating what are commonly called 'design freezes'). He argued that, in volatile markets, the opportunity to be flexible in the use of design technology provides considerable competitive advantages.

INNOVATION ACROSS-THE-BOARD: THE CASE OF BENETTON

So far in this chapter we have examined some of the features leading to good process performance in NPD. We shall conclude with a short case study that illustrates the importance of innovation not only in new products, but in all activities of the firm. As Rothwell (1992) says of his studies of the innovation process, it is not enough to be brilliant at one or two areas of activity; innovation depends upon doing all of them competently.

Benetton, the Italian clothing company, provides an example of a firm that is innovative across the range of its activities: from design and production to distribution, sales, and organizational structure.[3] Although the company is primarily a clothing company, it has diversified into cosmetics, glasses, luggage, shoes, and watches.

Benetton has enjoyed remarkable and sustained growth throughout the 1980s and 1990s. It is a family company with 82 per cent of shares family owned. The company's history began in 1957, when Luciano Benetton, a shop assistant in the textile industry, began to work with his sister, Guiliana, a knitwear factory worker. Luciano 'moonlighted' as a salesman for knitwear, and collected orders for Guiliana to design and make at home. By 1965 these activities had reached such a scale that they established a factory in Ponzano, which employed sixty people. Their two brothers, Gilberto and Carlo, assisted the company with administration, finance, and production. In 1968 the company opened its first shop in Italy, and by 1975 it controlled over 200 shops. In 1978 it internationalized, and by 1985 it controlled over 2,000 shops. Four years later it controlled 5,900 shops in eighty-two countries and by 1995 this had increased to 7,000 shops in 163 countries.

Innovation played a central role in Benetton's extraordinary growth. First, there was innovation in design. Benetton was one of the first companies to introduce pastel colours in clothing. When it did so, in 1965, it immediately hit upon a huge demand. The company was also an early entrant into the use of CAD of clothing (in 1980). This allowed greater ease and control of the design process and by 1996 Benetton was producing 4,000 design models annually from its CAD system.

Secondly, Benetton demonstrated innovation in production. In its first factory, Benetton used old knitting machines that had been modified to produce substantial productivity improvements. Its subsequent use of innovative

[3] This discussion owes much to the work of Fiorenza Belussi (see Belussi 1989).

production machinery included: knitwear stretching equipment (1972), automated knitting (1979), air-lift work tables and automated dyeing (1982). In 1993 and 1995 it built two new robotized factories near Treviso, Italy, capable of producing 120 million items of clothing a year. These factories, and the co-located distribution centre, are connected by a fibre-optic cable network, which includes the company's suppliers. Benetton also introduced a system of post-manufacture dyeing, so that the colours could be chosen after the clothes had been made, rather than the previous practice of assembling already dyed clothes. In an industry where preferred colours change seasonally, this allowed a much quicker response to customer demand.

Thirdly, in addition to innovation in the technology of production, Benetton is innovative in its organization of production. It has twelve factories, which, with the company's headquarters, employ around 12,000 people. It also subcontracts a subtantial proportion of its production work to some 500 small firms locally, employing around 20,000 people. Subcontracting accounts for 40 per cent of knitting and weaving, 80 per cent of cutting and grading, and 60 per cent of ironing and assembly. There are four categories of subcontractor: firms that are financially controlled by Benetton, affiliate firms, independent firms, and homeworkers. The advantages of subcontracting include the use of external managerial resources and significant reductions in labour costs.

Fourthly, the company's distribution system is innovative, involving a combination of automation and IT with a system of agents and franchising. At the centre of distribution operations are the Benetton shops, which are described as the 'antennae of the Benetton information system'. The shops report weekly sales figures and can, on this basis, determine micro-level trends. Initially, this system was centralized in company headquarters. When the system grew too large, it was decentralized to satellite electronic data processing (EDP) centres, where information from about 200 shops in each of four strategic markets generated the data to enable market trends to be forecast.

In 1997 Benetton signed an agreement with IBM to provide electronic point-of-sale (EPOS) terminals, software, and support for all 7,000 shops. According to Benetton's director of computer systems, this will eventually deliver common management methods in the shops. Operational management will be simplified and IT services will be distributed from the central data processing system to the point of sale. Benetton's information system produces a just-in-time response to consumer demand, enabling Benetton to get a new product to market six to eight weeks before its competitors and to respond within eight days to reorders from both domestic and foreign shops. This reduces inventory and the time capital is tied up in warehouses. The company's distribution centre is highly automated—with a staff of twenty, it dispatches 30,000 parcels daily.

The company's agents also play a central role in the Benetton system. There are around eighty agents, each responsible for a geographical area. Their functions include operating the distribution network, optimizing and promoting growth, communicating market trends, providing suggestions and new ideas, and giving technical support. Many of these agents are also shop-owners, and operate within Benetton's strict retailing policy. Benetton's shops are franchised and they are allowed to sell only Benetton stock (Benetton only sells to Benetton shops). The company defines shop styles, layout, and sales strategy. In the mid-1990s Benetton developed a preference for larger stores offering a wide range of Benetton goods.

There are many advantages of this retailing system to Benetton, the family company. The system guarantees outlets for Benetton products and hence reduces risk and uncertainty; it enables direct knowledge of customer preferences; it allows for easier and more efficient planning of production processes; and it allows the shops to be used exclusively for Benetton advertising and marketing. Benetton's advertising is notoriously controversial, but its image as an exciting and innovative company is enhanced by having its own volleyball, basketball, rugby, and Formula 1 racing car teams, a creative journal, *Colours*, and its own research school, Fabrica, for exploration and production in the arts.

According to Benetton, all these factors in combination provide the key to the company's success. A company report/promotion describes it as follows.

> There is a key to Benetton's success: the possession of a peculiar technology. Not technology in the traditional sense, it is different from 'revolutionary' techniques or unusually sophisticated machines . . . Instead, it is about a specific kind of know-how that the Benetton group feels it has acquired. This cannot be acquired in any way except by direct experience . . . This know-how enables the company to enter any budding market, develop it and reap the benefits of its growth.

The Benetton system does have its tensions and shortcomings. Suppliers are 'locked in' to its system and this limits their potential growth. Risk is pushed onto the shop-owners, and this is compounded by the Benetton policy of no returns of stock. Benetton has never succeeded in the United States, where there are regular complaints about its comparatively high prices, and dissatisfaction with its returns policy that goods have to be returned to the store of purchase. It continues to be very dependent upon sales in Italy, which is the source of 75 per cent of company profits. It is under continual scrutiny as a result of its confrontational advertising and, importantly, the Benetton system of production and distribution is increasingly emulated by its competitors.

None the less, Benetton has been phenomenally successful and the key to this success has been innovation across-the-board. Although technological innovations historically have been very important in both design and production and in the information system the company uses, these are combined with the organizational innovations in its production networks, franchising, and agents to create the special 'know-how' that the company defines as its core competence.

CONCLUSIONS AND SUMMARY

There is a great deal of research on the importance and process of new product development. In this area of MTI, at least, it is relatively easy to be prescriptive.

Products can be new to the world and new to the firm. They often take the form of incremental improvements to existing products. Firms can achieve competitive advantages not only through producing what customers want, but also, on occasion, through producing what they do not yet know they want. There are many benefits of NPD, and these extend beyond the simply financial. Assessing NPD encompasses both efficiency and effectiveness factors: how good are firms at developing new products, and are firms producing the right products? The latter is a strategic management issue concerned with whether the firm is cumulatively building its competencies (examined in Chapter 6), and whether the intellectual property produced is protectable (examined in Chapter 8). Successful firms, like Benetton, are not only innovative in their NPD, but are innovative in all their activities.

Factors that encourage efficient NPD include: internal organizational integration, particularly between R & D, marketing, and manufacturing; human resource management, particularly concerning project managers; and the use of project management techniques, such as stage-gate systems. Many of the internal organization practices of Japanese firms have been discussed, and the important role of job rotation in encouraging good communications between all those involved in NPD has been described. Many of the best practices in rewarding and providing incentives for NPD teams, such as opportunities for stock options and spin-outs, can be found in Western firms. Many of the practices of NPD in manufacturing industry also apply to services, as seen in the case of Citibank.

Despite all that is known about NPD, there are still many instances of the failure of new products. These must be seen as learning opportunities so that the reasons for the failure are not replicated.

CHAPTER

5

The Management of Operations and Production

WHAT ARE OPERATIONS AND PRODUCTION AND WHY ARE THEY IMPORTANT?

Operations can broadly be defined to include all activities related to transforming inputs into final outputs (Porter 1985). The process by which the transformation occurs has to be managed so as to add value by making something useful. Operations include issues such as inventory and workflow control, quality management processes, and supply chain management. Operations are, therefore, important to a wide range of companies, both 'services' and 'manufacturing' (bearing in mind the distinction between these two are increasingly indistinct).

Production is more narrowly defined as including all the activities involved in the physical transformation of artefacts.[1] Production includes assembly and testing, and processes such as machining, extruding and forging, and automation. Production is important in a wide range of industries and encompasses a broad range of activities, including, for example:

- *heavy industry*, such as foundry work, where molten metals are cast; large-scale assembly, such as car plants where pressing machines stamp out car bodies, and robots move, weld, and paint; and continuous process oil refineries and petrochemical works;
- *light industry*, such as the manufacture of furniture, clothing, and food; 'white goods', such as refrigerators, ovens, air-conditioners; consumer electronics and computer assembly;
- *high-technology industry*, which makes and assembles high-technology

[1] Operations and production are sometimes conflated into the term 'manufacturing'. In this chapter, where work is cited that refers to manufacturing, this conflation will be retained. The purpose in separating operations and production elsewhere is to emphasize the ubiquitous role of operations in business—they include service as well as manufacturing activities—and hence avoid restricting the discussion to manufacturing industry.

products such as semiconductors and produces complex pharmaceutical products;
- *precision engineering,* which machines high-value metals to tolerances measured in microns in, for example, the aerospace industry;
- *specialist production,* such as the making of one-off specialist projects, such as air traffic control systems; and products and equipment, such as prefabricated houses, or sound systems for touring rock bands.

In innovative firms, the operations and production domain is also involved in the design of the products and components it makes, and influences the company's overall strategy. In manufacturing companies, operations and production activities utilize the majority of capital assets (estimated at between 80 and 85 per cent), and can account for 80 per cent of company costs (Brown 1996). They are, therefore, a critical determinant of the price of a company's products. They also affect important non-price factors in competitiveness, such as quality, delivery, reliability, and responsiveness, and the ability to offer variety and to customize. In a study of twenty-three major development projects at eleven US and European pharmaceutical companies, Pisano and Wheelwright (1995) found that problems in manufacturing processes either delayed a new product launch or inhibited its commercial success once it was on the market. The operations and production domain, therefore, can provide the basis of firms' ability to compete.

The ways firms make their products are changing profoundly. In Japan, the role of the factory has changed so much that it has led one analyst to refer to the 'factory as a laboratory' (Baba 1989). This reflects a shift towards production as an area that provides a major focus for technological learning (Pisano 1996). As Kenney and Florida (1993: 65) put it, 'the factory is no longer a place of dirty floors and smoking machines, but rather an environment of ongoing experimentation and continuous innovation'. The ultimate aim of these innovations in operations and production is to allow firms to meet market requirements quickly and cost effectively. They are aimed at producing 'mass-customization' products that are innovative, capable of being customized to meet specific requirements, and economical to produce (Kotha 1995).

As with other areas of MTI, issues of work organization and the increasing knowledge content of work are centrally important. Similarly, the many techniques that can be used in operations and production management are most effective when they are informed by a strategy. The focus of this chapter will be primarily on the strategic management of operations and production, although some of the techniques of management will be discussed. Our concern here will be to understand the role of, and strategic opportunities provided by, operations and production and the ways firms can quickly and reliably produce high-quality innovative products and services whilst maintaining the cost advantages previously achieved through large-scale production for mass markets. The techniques described are applicable to companies in a wide range of business and of all sizes, including small firms.

SOME TECHNIQUES OF OPERATIONS AND PRODUCTION MANAGEMENT

Operations management includes a number of techniques such as total quality management (TQM), and the various generations of workflow control systems ranging from materials requirement planning (MRP) to manufacturing resource planning (MRPII), to enterprise resource planning (ERP).

Quality management

Quality is a major strategic issue, as guarantee of quality provides distinctive competitive advantages. Excellent quality in products and services is an essential element of a firm's reputation. Poor quality may lead to the costly reworking and recalling of products. For example, Ford had to recall over 2.7 million cars and pick-ups between 1986 and 1993 because of defects (Brown 1996).

Various techniques are used to improve quality. Quality control (QC) techniques provide statistical tools for identifying trends and quality metrics. Quality assurance (QA) develops procedures and policies to ensure repetitive, high-quality processes. QA often utilizes international quality standards, such as the International Standards Organization (ISO) 9000 series. ISO 9000 essentially states that an organization should have a QA procedure and that it is up to the organization to identify those procedures and put them in place. To obtain ISO 9000 certification firms must develop detailed procedures for ensuring quality at all stages of production (for the whole of the production process, not just individual products) and produce strict documentation of adherence. Many firms insist that their suppliers meet these quality standards.

Total quality management (TQM) is a philosophy of continuous improvement. Insurance companies were amongst the earliest adopters of TQM, and now it is used widely in business, from financial services to metal-bashing manufacturing. TQM often puts the responsibility for quality on each employee, who is expected continuously and proactively to assist in overcoming any difficulties in meeting the standards expected by customers. It involves detailed consideration of organizational matters, and requires significant investment in training. The increasing strategic significance of quality is illustrated in Fig. 5.1. At the turn of the century, quality issues in production are often taken as given. Every firm is expected to be able to produce and deliver services to very high quality standards.

TQM has mistakenly been seen as a simple technique for improving production quality, rather than a broader philosophy affecting all the activities of the firm. TQM, like many of the techniques that will be discussed, has suffered a great deal from its over-promotion, and its perceived status as a management fad. Cases of unsuccessful introduction and detrimental effects have been cited as evidence of its lack of efficacy. However, in one of the most detailed analyses of TQM, Cole (1995) found that it could be valuable, provided there is

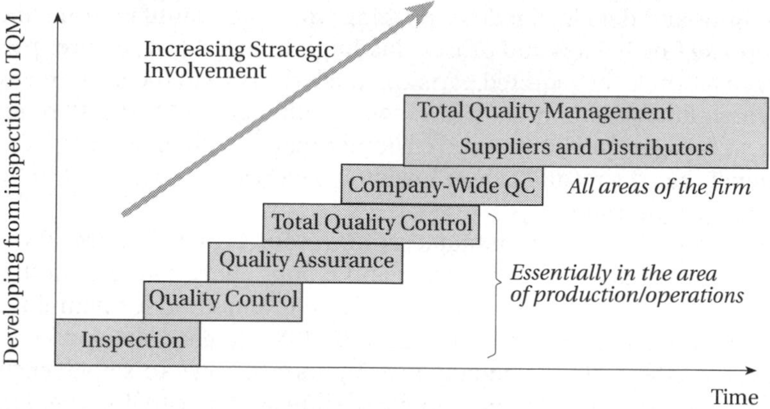

Fig. 5.1. The evolution of quality management
Source: Brown (1996).

- sustained top-management leadership for the quality initiative and active application to their own management activities;
- relentless focus on the customer, both in setting strategic objectives and building organizational routines that integrate the firm internally and externally so as to identify and meet customer needs;
- decentralization of decision-making responsibility to a well-trained problem solving labour force (i.e. employee participation in decision-making);
- attention to reducing organizational barriers between departments and levels so that cross-functional management becomes normal operating procedure;
- combined emphasis on incremental continuous improvement and break-through strategies;
- realignment of reward and measurement systems, both formal and informal, to support these new directions;
- decentralized information and decision support system that provides data to a broad set of personnel.

Workflow and materials planning

The holding of stock (or inventory) of raw materials, part-processed products, and final products has traditionally acted as a buffer for the problem of changes in demand and supply in operations and production; however, with so much capital tied up non-productively in inventory, there are incentives to reduce stock holdings and conduct operations with 'zero-buffers'. Materials requirement planning (MRP) is a computerized production scheduling system that emerged in the 1970s and details schedules pushing work into manufacturing lines with appropriate finish dates and process loadings. In this way it requires manufacturing to produce the required parts and push them on to the next process until they reach final assembly stage (Hill 1985). Although an MRP system relies on forecasts being largely correct and customer schedule changes being restricted, it generally has the result of lower inventory levels and things being at the right place at the right time.

MRP evolved first into manufacturing resource planning (MRPII), which includes the capacity to plan tooling and labour allocation and, in the 1990s, into enterprise resource planning (ERP). ERP is a method for planning, tracking inventory, and production. The aim of ERP is to generate and use data to assess how efficiently a company is using its resources to satisfy customers. SAP's R/3 product provides an example of software that can link every part of a company's operations (see Fig. 5.2). Its use has allowed Monsanto to cut production planning from six to three weeks, trim inventories, reduce working capital, and increase its bargaining power with suppliers. It is estimated to save Monsanto $200 million annually (*Business Week*, 3 Nov. 1997).

These various techniques, starting with MRP in the 1970s following on to ERP by the 1990s, are sometimes seen as 'magic bullets'—ways of 'solving' complex operational problems. However, as we shall see, although the techniques can undoubtedly be useful if managed well, the challenges of effective operations and production are often to do with organization and skills, and managing them strategically. Unless these are addressed, the techniques will have little benefit.

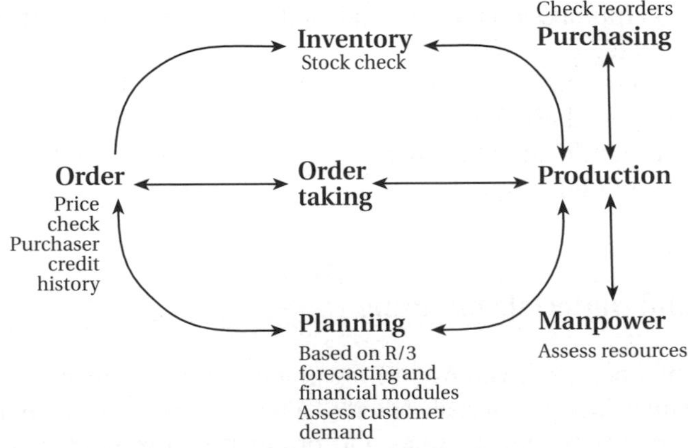

Fig. 5.2. SAP's enterprise application software

LEAN PRODUCTION

Lean production is a term that derives from the Formula 1 racing-car industry, and was used to explain how engines were developed quickly and efficiently by eliminating wasted time and effort. It was used in a worldwide study of the auto industry that contrasted the use of lean production methods found at that time in Japan with the mass production methods found in the United States and Europe (Womack *et al.* 1990). The differences in performance were immense, with lean production using half the hours of mass production methods to design and make a car, in half the factory space, with half the labour, and half the investment in production tools. Using lean production methods, car-makers also produced greater product variety and higher quality.

Lean production includes just-in-time and supply chain management techniques. Just-in-time (JIT) is a managerial technique whose aim is to reduce inventory. It originated in US supermarkets, which managed their systems so that foods were delivered just-in-time to be sold. This meant goods did not have to be stored by the supermarket, thereby reducing storage requirements and the amount of capital invested in unproductive inventory. JIT methods were successfully transferred to Japanese manufacturing firms, which produced and delivered finished goods just-in-time to be sold, sub-assemblies just-in-time to be assembled into finished goods, fabricated parts just-in-time to go into sub-assemblies, and purchased materials just-in-time to be transformed into fabricated parts (Schonberger 1983). As shown in the supermarkets, JIT is a 'pull' system, with demand pulling components and parts to where they are needed. It prevents materials being inactive in the system, thereby avoiding unwarranted costs of inventory. JIT is now widely used in manufacturing and service companies internationally.

Supply chain management is the other feature of lean production. Traditional customer–supplier relationships in the United States and much of Europe have been 'arm's length' and based on a form of 'spot-trading', where tenders for particular work were put out to a number of potential suppliers. The contract was usually awarded on the basis of lowest cost and there was no assumption of continuity in the relationship. Based to a large extent on Japanese and (to some extent) Scandinavian experiences, far greater levels of integration have evolved between suppliers and customers, and a general trend has emerged amongst larger firms towards operating with a few 'primary' subcontractors, who have responsibility for coordinating 'secondary' subcontractors. The extent of the rationalization process in the number of suppliers in the United States is shown in Table 5.1.

In contrast to the arm's-length structure traditionally prevalent in Western countries, Nishiguchi (1994: 122) found a *clustered control* structure in Japan in which

> firms at the top of the clustered control structure buy complete assemblies and systems components from a concentrated base (and therefore relatively limited number) of first-tier

TABLE 5.1. *Rationalization process for US suppliers*

Firm	Number after supplier consolidation	Before supplier consolidation	Change (%)
Motorola	3,000	10,000	−70
DEC	3,000	9,000	−67
Ford United States	1,000	1,800	−44
GM United States	5,500	10,000	−45
AT&T	240	1,471	−84

Source: Brown (1996).

subcontractors, who buy specialized parts from a cluster of second-tier subcontractors, who buy discrete parts or labor from third-tier subcontractors, and so on . . . this system absolved those on top of the hierarchy from the increasingly complex controlling functions typical of external manufacturing organizations.

According to Nishiguchi, this arrangement has had a marked impact on the efficiency of the system. He argues that

> Japanese automotive and electronics producers have achieved notable growth not only by unilaterally exploiting subcontractors but by strategically creating, and benefiting from, distinctive institutional arrangements in subcontracting based on problem solving. These new arrangements institutionalized the goal of continuous improvement with the aid of systematic checking mechanisms . . . Prime contractors benefit from the subcontractors' enhanced performance, and the result is better design, higher quality, lower cost, and timely delivery. At the same time, the establishment of rules to share fairly the profits from collaborative design and manufacturing has encouraged the subcontractors' entrepreneurship and their own symbiotic relationships with their customers. Benefits from the subcontractors' commitments have reached their customers as well; thus a virtuous circle has emerged. (Nishiguchi 1994: 6)

Subcontractors benefit from this system through the existence of stable contractual relations, which are usually automatically renewed, improved technological learning, which occurs as customers make considerable efforts to upgrade the equipment and skills of their suppliers, and improved growth opportunities.

The extent of the cooperation between contractors and subcontractors extends to cost information, whereby customers frequently request detailed cost data from subcontractors, opening the way to rational price determination (Sako 1992). The extent of the electronic integration of suppliers in the US car industry is seen in the Automotive Network Exchange, which is a transmission control protocol/Internet protocol. This integrates suppliers with Ford, GM, and Chrysler (and suppliers with suppliers) in standard protocols, and enables exchanges of product engineering data and EDI, and has e-commerce and e-mail applications. Any company wishing to sell the 'Big 3' has to be a party to this computerized system, and be prepared to bargain over prices on line.

Contractors and subcontractors seek to reduce costs by means of joint problem solving. One mechanism for this is the use of resident engineers. In an arrangement pioneered by automotive companies, engineers from supplier companies join in the development and manufacturing activities of the customers and, according to Nishiguchi (1994), have had an important role to play in promoting the benefits of bilateral design and collaborative manufacturing.

Decisions to outsource or subcontract have great strategic significance. Once capabilities are bought in rather than kept in-house, those capabilities may be lost to the firm forever. Consequently, it is important to have strong links through high-trust relationships with suppliers. There is compelling evidence of the advantages to be found in having close, obligational contracting relationships between firms (Sako 1992), and these issues are examined in greater detail in Chapter 7.

A brief note on benchmarking

Benchmarking or comparing performances between factories and firms is common practice. Firms have become experienced at choosing comparators against which to measure their performance. Although there are many difficulties associated with the comparisons, careful collection of data and their cautious interpretation can be useful in identifying weaknesses in performance and establishing targets for improvement. It is mistaken to assume that the aim of improvements resulting from the benchmarking exercise is to meet the levels of performance achieved by best-practice firms. If the exercise is to be utilized fully, the performance has to exceed the likely improvements of those best-practice firms. As shown in Fig. 5.3, the aim of company A *should*

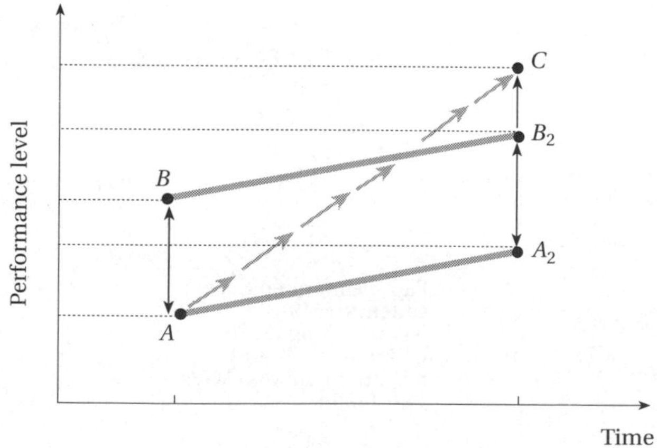

Fig. 5.3. Targeting improvement through benchmarking

not be to continue its improvement trajectory to A2, *nor to emulate that of company* B *and its improvement to* B2, *but instead to aim for* C.

AUTOMATION

Automation can take many forms, ranging from the automation of existing activities, such as design or machining or coordination, to the pinnacle of the automated factory, computer integrated manufacture (CIM). A good way of conceptualizing automation is to use the approach of Kaplinsky (1984), who

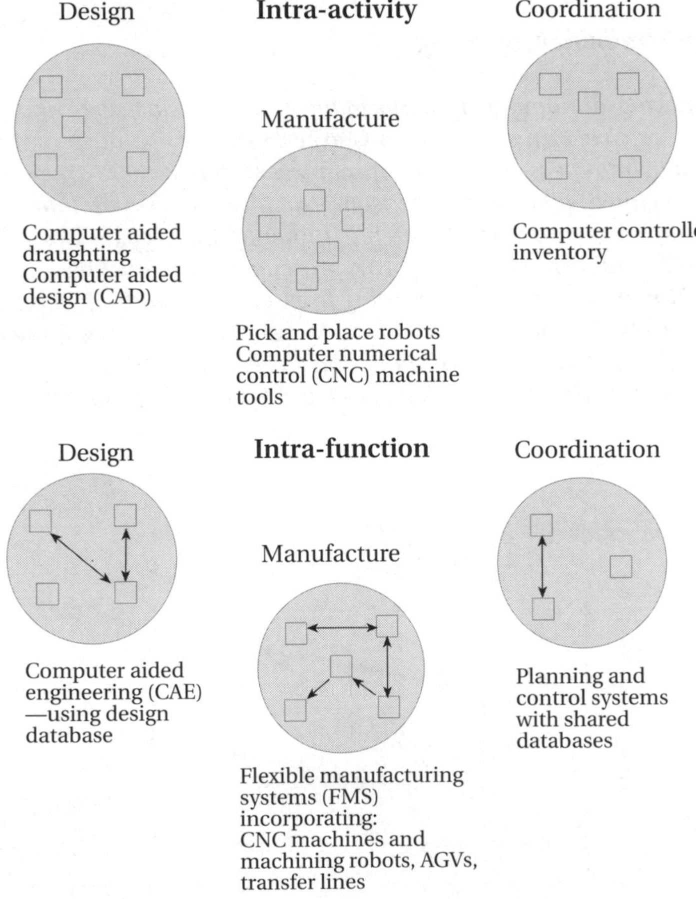

Fig. 5.4. Forms of automation
Source: based on Kaplinsky (1984).

Inter-function

(a) Computer aided design/computer aided manufacturing

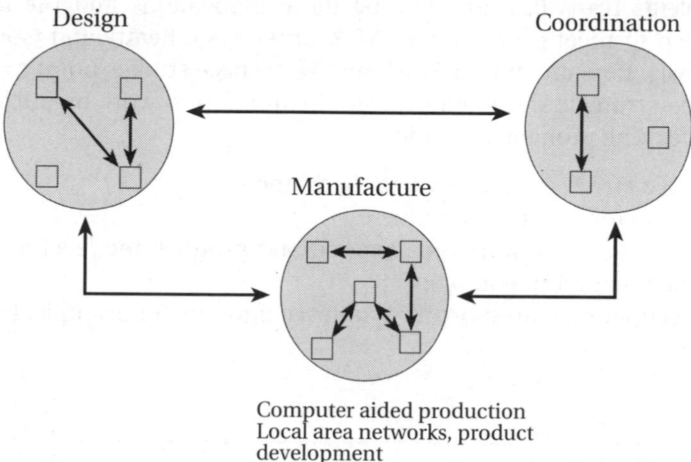

Computer aided production
Local area networks, product development

(b) Computer integrated manufacturing

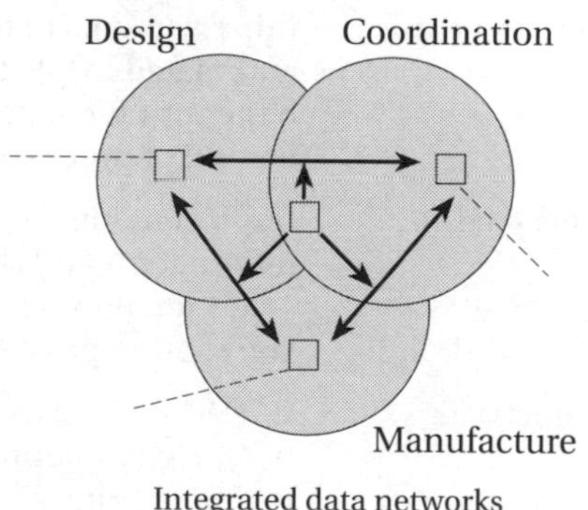

Integrated data networks
Enterprise resource planning

Fig. 5.4. *Continued.*

describes three forms of automation: intra-activity, intra-function, and inter-function (see Fig. 5.4). Automation may also extend beyond the firm and integrate the firm's activities with those of others. It uses EDI for the transfer of information, and increasingly e-commerce on the Internet to facilitate inter-company transactions processing systems.

Automation, in the form of computer-integrated manufacturing, has often been considered to provide the answers to all the problems of operations and can have the potential to help over many of the difficulties encountered (see Box 5.1); however, the promises of automation are rarely met (Voss 1994; Upton 1995). Over the years there have been a number of studies that have revealed the problems firms face in adopting these innovations and the failures of automation to meet expectations (AT Kearney 1989; Beatty and Gordon 1991; Bessant 1993; Bessant and Buckingham 1993). These studies point to technical difficulties, strategic shortcomings, and inattention to work organization and skills issues. The problems include:

- lack of a strategic framework for investments;
- lack of planning and foresight;
- inability to adapt work organization and produce the level of organizational integration required;
- preoccupation with short-term returns, through, for example, labour cost savings;

Box 5.1. ***CIM a solution?***

Main problem issues	Potential contributions offered by CIM
Producing high-quality standards	Improvements in overall quality via automated inspection and testing, better production and more accurate control of processes
High and rising overhead costs	Improvements in production information and shorter lead times, smoother flow; less need for supervision progress chains
Introducing new products on schedule	CAD/CAM shortens design lead time. Tighter control and flexible manufacturing smooths flow through plant and cuts door-to-door time
Poor sales forecasts	More responsive system can react quicker to information fluctuations
Inability to deliver on time	Smoother and more predictable flow through design and possible accurate delivery
Long production lead times	Flexible manufacturing techniques reduce set-up times and other interruptions so that products flow smoothly and faster through plant

Source: based on the work of John Bessant.

Box 5.2. ***Success and failure in implementing automation***

Success is associated with	Failure is associated with
Top management commitment at all stages of the project	Lack of commitment
Clear strategic vision, communicated throughout the organization	Lack of clear strategy and/or its effective communication to the rest of the organization
Shared views of project aims and implementation approach	Lack of shared view and unresolved conflicts regarding design and implementation
Multi-function project teams, multi-function perspective	Single function teams, unilateral perspective
Effective conflict resolution within team	Unresolved conflict over key implementations issues
Extensive user education to give understanding of broader implications and purpose of system	Minimal training for operation
User involvement in system design (of hardware/software, jobs, structures, roles, etc.)	Unilateral design, organization expected to adapt to systems rather than change system
Close involvement with suppliers	Minimal involvement
Readiness to re-examine and change existing procedures	Attempt to computerize what is already there
Performance measures that reflect broader organizational effectiveness	Performance measures narrowly defined and related to efficiency at local level
Flexibility in design and continuous monitoring to adapt to unexpected changes	Inflexibility in system in response to unexpected changes in environment

Sources: Bessant (1991); Bessant and Buckingham (1993).

- failure to realize the advantages of systemic integration;
- the technological complexity of CIM;
- the need to undergo extensive learning and adaptation;
- a requirement for greater breadth of skills and flexibility in the workforce.

An examination of the success and failure factors in the introduction and use of automation shows that these are similar to those found in new product development. They are shown in Box 5.2. Some of the major factors that affect the uptake and use of automation and operations and production management techniques include investment appraisal techniques, national approaches to operations and production management, the extent of organizational integration (both internal and external), and the question of strategy. These will now be examined in turn.

INVESTMENT APPRAISAL TECHNIQUES

Firms tend to use simple quantifiable financial indicators, such as *payback periods* and *discounted cash flows,* to assess their operations and production investments. These tend to create risk-averse and short-term attitudes. The payback-period method is simple and widely used. A company invests only in those projects that can pay back the original investment in a certain number of years, usually two or three years. This method ignores returns post payback period, and minimizes risk.

The discounted cash flow (DCF) enables interest rates to be included in investment appraisals. When an investment takes a long time to pay-off, the company pays more interest to finance it. DCF is a means of allowing for this in deciding whether to invest. The returns a company expects to make in the future (for example, in five years) are reduced in its calculations to allow for the fact that the company will have to pay interest on that part of the investment for a full five years. That part of the investment that comes back to the company quickly (for example, in one year) is more valuable to the company because it does not have to pay as much interest before it gets its money back. Since returns a long way in the future are 'heavily discounted', the DCF method favours investments that give a quick return (although not to the same extent as the payback-period method, which does not count returns after the period).

Even the commonly used return-on-investment (ROI) indicator has problems in respect of operations and production. In addition to encouraging short-term investments,

- it is very sensitive to depreciation write-off variances, which makes comparisons between facilities using different approaches to depreciation unfair;
- it is sensitive to book value and older facilities can therefore enhance ROI artificially;
- it fails to incorporate non-quantifiable information, such as competitor behaviour or the specifics of product markets.

Therefore, according to Brown (1996: 53) 'decisions concerning manufacturing investment, in terms of new processes or technology, have to be made on long-term manufacturing advantage and not narrow financial criteria such as ROI or immediate cost savings'. As with R & D investment, an 'options'-based approach to investment in operations and production has advantages. Hayes and Pisano (1994: 79) argue that,

> according to the new approach to manufacturing strategy, managers should think about investments more in terms of their capacity to build new capabilities. Rarely, if ever, is a strategically worthwhile capability created through a one-shot investment. Capabilities that provide enduring sources of competitive advantage are usually built over time through a series of investments in facilities, human capital and knowledge . . . Investments can create opportunities for learning. These opportunities are a lot like financial options: they have value, and that value increases as the future becomes more unpredictable.

An important corollary of this approach is that investment in factories and facilities are different from many other investments in the sense that they are 'organic'—that is, they take time to grow and involve various integrated elements such as machines, people, and organization. They are not investments that can be readily switched on and off. Once the 'organism' is dead it cannot quickly be revived. Hayes and Pisano argue that the capabilities embodied in human capital are analagous to human muscles, which atrophy with disuse and can deteriorate irreversibly.

NATIONAL DIFFERENCES IN APPROACHES TO PRODUCTION

There are broad international differences in productivity growth in manufacturing industry. For the period 1985–95, for example, average annual growth rates in manufacturing output in the OECD ranged from 3 per cent in Finland to –0.3 per cent in Greece. The figures for the United States and Japan were 2.2 per cent and 2.5 per cent respectively.[2] These differences in productivity growth are in part explained by the different approaches to the use of manufacturing technologies and operations management.

There are enormous differences between firms' abilities to manufacture. Womack *et al.* (1990) compared assembly plant efficiency in the auto industry by measuring the hours required to produce a vehicle. There were important variations in performance between and within countries. For example, the best assembly plants were nearly eight times more productive than the worst (they could make a car in ten hours compared to eighty). Some Japanese car firms manufacturing in Japan were the most productive internationally, but the best US and European firms operating in their home countries were more efficient than the least efficient Japanese firms.

Some of the national differences are shown in Box 5.3, which summarizes the different approaches taken by Japanese and British firms to investment in assembly robots. There are differences in the motivations for, views on the role

[2] Measuring productivity growth is fraught with difficulties; see Lester (1998) for a discussion of the problems faced.

Box 5.3. ***Assembly robots in the United Kingdom and Japan***

Technology issue	UK	Japan
Primary motives for development and and adoption	To increase productivity and improve quality through the elimination of direct labour	To improve flexibility of production but continue to reduce costs through the elimination of waste
Technological trajectory pursued	Complex, sophisticated technology consistent with long-term goal of CIM, a computer systems approach	Relatively simple proven technology with continued reliance on operators; essentially a production engineering approach
Manufacturing	Reduction in diversity of production to facilitate further automation and computer integration	Flexible, but low-cost production
Source of most significant developments	Specialist suppliers essentially 'technology push'	Major uses essentially 'demand pull'

Source: Tidd (1991).

and aims of, and sources of technology in both countries. The Japanese rather than the British approach provides the greatest competitive advantage (Tidd 1991).

In a study of the use of technology in Lucas and Bosch, a British and a German firm working in the same industry, national differences in human resources, and in the education and training systems, were found to affect the use of technology. Bosch, the German company, had a distinct advantage in the way it introduced new technology into its generally better trained workforce. It was able to introduce changes continuously and productively, compared with Lucas, where changes appeared discontinuously and disruptively (Harding 1995).

Two countries that are renowned for their production skills and prowess are Germany and Japan, and some of the characteristics of production in these countries that have led to their success are outlined in Box 5.4.

A brief note on the role of the engineer in industry

In Japan and Germany the engineering profession is highly regarded. In contrast, a key problem for many firms elsewhere is the lack of importance

Box 5.4. ***The German and Japanese systems of production***

German	Japanese
Meticulousness	Cleanliness and orderliness
Technical strength throughout the managerial hierarchy	Broad task range for workers including maintenance of machines
High and broad skill levels in employees	Avoidance of inventory at all stages of production
Strong customer orientation	High-quality consciousness
High awareness of competition	A lifetime mutual commitment
Long-term orientation	Participative management and emphasis on teamwork
Good worker–manager relationships	Design and production of own production equipment
Consensus-seeking decision processes	Constant attempts to improve
Building on existing strengths: incremental innovation	Long-term relationships with a limited number of suppliers

ascribed to manufacturing, as seen in the limited number of its representatives on boards of directors. According to Wickham Skinner (1985: 214), 'to many executives, manufacturing and the production function is a necessary nuisance—it soaks up capital in facilities and inventories, it resists changes in products and schedules, its quality is never as good as it should be and its people are unsophisticated, tedious, detail-oriented, and unexciting'. In such companies, manufacturing is seen as 'cost' (as opposed to marketing, which is seen as income generating) and as a 'problem' to be solved.

THE INTERNAL INTEGRATION OF OPERATIONS AND PRODUCTION

Perhaps the major problem with the introduction of new operations and production technologies is the inability to adapt the organization to the opportunities the technology provides. The extent of the need for aligning technological and organizational change is shown in Fig. 5.5. The higher the level of technology, moving from intra-activity to inter-function automation, the greater the need for organizational change. As Voss (1994) argues, the challenge lies in the implementation of manufacturing technology change.

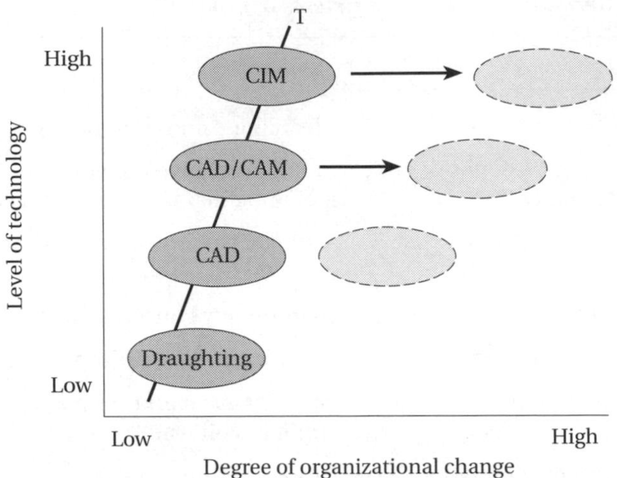

Fig. 5.5. Aligning technology and organizational change
Source: Voss (1994).

The importance of work organization issues is seen in the case of computer-integrated manufacturing (CIM). CIM is often seen as a means of improving the flexibility of the production domain. Flexibility is not an easy concept and can mean different things to different people, but, according to Upton (1995), it involves:

- *increasing product range*: the ability to make a small number of different products, or a large number of slightly different products;
- *mobility*: the plant's ability to change nimbly from one product to another, minimizing response times and inventory;
- *uniformity*:the capacity to perform well when making any product within a specified range.

As such, flexibility is an important source of value enhancement from a given capital stock.

In Upton's study of sixty-one paper manufacturers in the United States, he found that the flexibility of the plants depended much more on people than on any technical factor. 'Although high levels of computer integration can provide critically needed advantages in quality and cost competitiveness . . . operational flexibility is determined primarily by a plant's operators and the extent to which managers cultivate, measure, and communicate with them' (Upton 1995: 75).

There are many ways of organizing work around a similar technology. The different forms of work organization used in three companies using similar computer numerically controlled machine tools in similar markets are shown

TABLE 5.2. *Work functions undertaken by the machine tool operator in three companies*

Company	Work planning	Programming	Program proving	Tool setting	Machine setting	Workpiece handling	Machine observation
A	XXX	XXXX	XXXX	XXXX	XXXX	XXXX	XXXX
B		XX	XX	XXXX	XXXX	XXXX	XXXX
C				XX	XX	XXXX	XXXX

Note: XXXX: always involved; XXX: frequently involved; XX: occasionally involved.
Source: Dodgson (1985).

in Table 5.2. In Company *A* the machine tool operator performed all the functions around the machine, including the 'clever bits' to do with computer-program writing. In Company *C*, by contrast, the operator did very little but attach parts to the machines and watch them operate.

A wide range of factors influences the way work is organized around particular technologies, and these need to be considered when introducing new technologies (see Fig. 5.6). Much depends on the choices made by managers. Although there are technical constraints (it does not pay to have a highly skilled worker capable of computer-programming watch the machine during the production of large batches), and trade unions might have an influence in determining who does what, managers generally have discretion in the choices

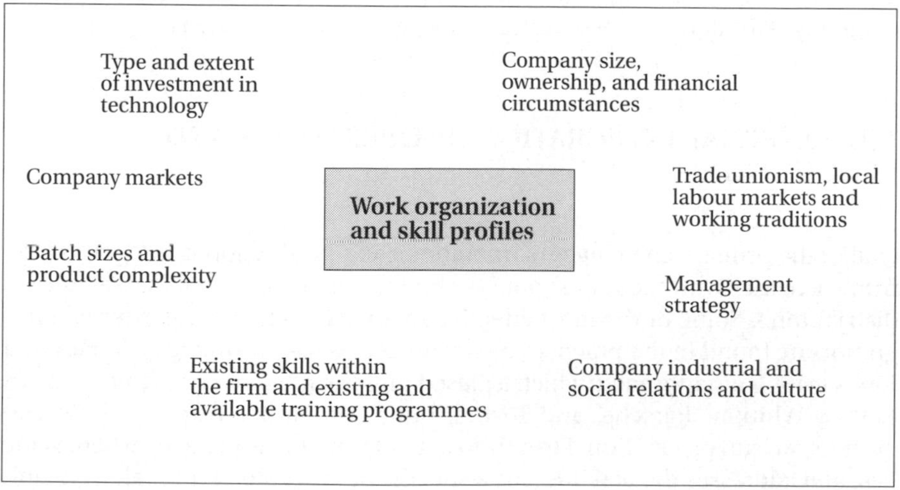

Fig. 5.6. Major influences on the way work is organized around new technology
Source: Dodgson (1985).

of work organization used. As we saw in the characteristics of production in Germany and Japan, the emphasis in these countries is on encouraging highly skilled workers to perform a range of functions so as to encourage flexibility. These approaches are supported and sustained by the training policies pursued in these countries (Harding 1995; Dore and Sako 1998). Such practices are also common in the United States and Europe.

Mobilizing the problem solving capabilities of employees is an essential part of policies for TQM and the continuous improvement in firms' operations. Incremental, continuous improvement is often the most important form of innovation in production activities (Bessant 1998). Giving greater autonomy and task discretion to the individual applies also to shop-floor teams. Quality circles, where workers from a particular work area meet to discuss quality and productivity issues, are one form of team structure that is very common in Japan. These have proved very advantageous to firms in dealing with operational matters, but their transfer to Western companies has not always been successful (Oliver and Wilkinson 1988). Harley Davidson, however, provides an example of a company that addressed quality problems through using TQM and devolving responsibility to shop-floor workers so as to give them autonomy over production decisions. By means of changes such as the creation of autonomous work groups, it replaced the number of product schedulers from twenty-seven to one (Brown 1996), and built a reputation as a reliable high-quality manufacturer.

As we saw in Chapter 2, many companies are changing from functional to process-based organizational structures. Business Process Re-engineering has enjoyed great success as a popular technique for this reorganization. It has not, however, produced noticeable productivity increases (Lester 1998), and has had many negative consequences, particularly job losses. The principle of organizing by process is valuable none the less, provided it takes account of existing company skills and routines, and is supported by appropriate IT.

THE EXTERNAL INTEGRATION OF OPERATIONS AND PRODUCTION

One of the primary challenges of operations and production is integrating the firm's activities with those of suppliers and customers (the final market and/or distributors). Some of the most effective methods of facilitating external integration are found in the practices characterized as *lean thinking* (Womack and Jones 1996). This approach, which is based on research into companies such as Pratt & Whitney, Porsche, and Toyota, builds on earlier work by these two authors on lean production. Lean thinking helps managers specify where value lies, and addresses the activities for a specific product along the 'value stream', incorporating the activities of a firm and those of its suppliers, customers, and distributors.

The integration of production activities occurs at two levels: within local production networks and within international networks. Local production networks—the concentration of many small and medium-sized firms in narrow regions with highly interdependent interactions among themselves and with parent firms—is an important aspect of industrial organization. This is seen in Japan (Itoh and Urata 1994; Whittaker 1997), and in Europe and the United States. Such small-firm networks play a central role in diffusing technologies, building skill bases, and achieving flexibility in production. This is seen particularly clearly in the success of the production networks in the electronics industry in Taiwan (Mathews 1996). Local networks have been discussed in Chapter 2 and we will consider this further in Chapter 7. Here the focus is on international production networks.

International production networks

There is a variety of explanations for why firms have production plants overseas. These include labour cost and other resource advantages, proximity to markets, maximizing returns to existing capabilities, developing new capabilities, and the desire to pursue global strategies. Multinational companies operate global supply chains, often locating more complex work in developed countries, and simple assembly-type operations in developing nations. Asia was the base of expanding international production networks throughout the 1990s (and was by far the largest recipient of direct foreign investment).

Samsung exemplifies the strategy of a company developing an international network within the Asian region. Fig. 5.7 shows the way Samsung created its production networks within Asia: Fig. 5.7*a* shows the historical path of development and Fig. 5.7*b* the level of integration between the parts of the network, as parts flow between countries. The integration of these networks allows firms like Samsung to achieve locational advantages through labour-cost differences and access to government incentives, and also provides great flexibility through the possibility of multiple sourcing.

Overseas investments assist local firms not only to improve their levels of operations and productions technology, but also to improve the quality of their management. In Indonesia, for example, the investment of Toyota in the local company PT Astra has enabled the local firm to improve its use of quality management techniques (Sato 1998).

There is empirical evidence that the various modern organization approaches and management techniques discussed so far can improve productivity. Fig. 5.8 is based on a large sample of capital equipment producers and it shows the productivity increases that occur when these techniques are used. However, these approaches and techniques (like investments in technological hardware) are effective only when they are directed by, and are part of, a manufacturing strategy. Womack and Jones (1996) argue that, while elements of lean production have been keenly adopted by industry, the overall philosophy or

(a) The historical path of Samsung's development

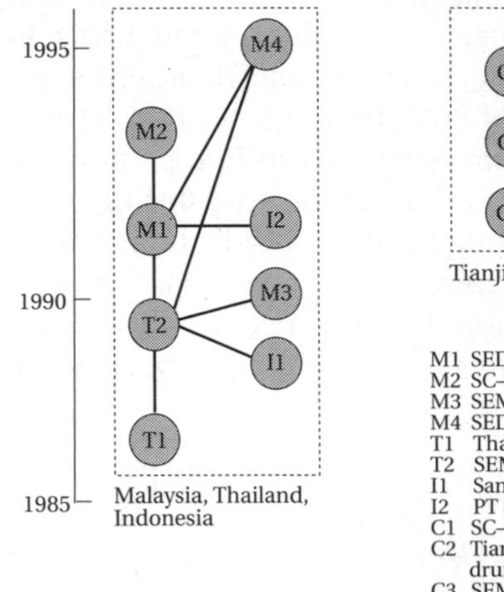

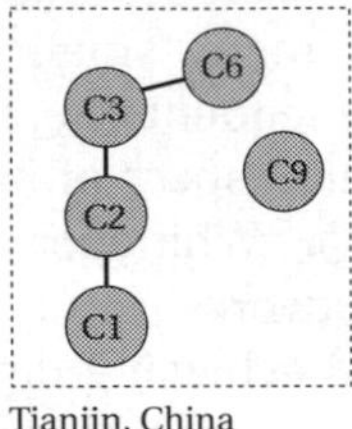

Tianjin, China

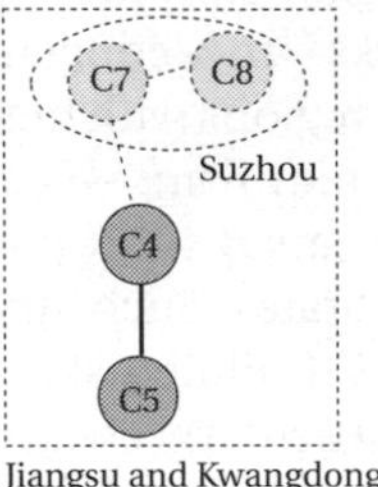

Jiangsu and Kwangdong, China

M1 SED–Malaysia (CPTs, 1991)
M2 SC–Malaysia (CRT glass bulbs, 1992)
M3 SEMA (microwave ovens, 1991)
M4 SEDM (colour monitors, 1995)
T1 Thai Samsung Electronics (CTVs,VCRs, and washing machines, 1988)
T2 SEM–Thailand (CTV and VCR components, 1990)
I1 Samsung Maspin Indonesia (refrigerators, 1989)
I2 PT Samsung Matrodata Electronics (VCRs and audio products, 1991)
C1 SC–Tianjin (rotary transformers, 1992)
C2 Tianjin Samsung Electronics Company (VCRs, VCR decks, and VCR drums, 1993)
C3 SEM–Tianjin (VCR drum motors. 1993)
C4 Huizhou SEC (audio products, 1992)
C5 SEM–Dongguan (speakers, keyboards, etc., 1990)
C6 Tianjin Tongguang Samsung Electronics Co. (CTVs, 1995)
C7 SSESC (semiconductors, 1995)
C8 Suzhou SEC (refrigerators, microwave ovens, washing machines, and air-conditioners, 1995)
C9 Samsung Aerospace Industries (cameras, 1994)

C China I Indonesia T Thailand M Malaysia

(b) The flow of components and parts

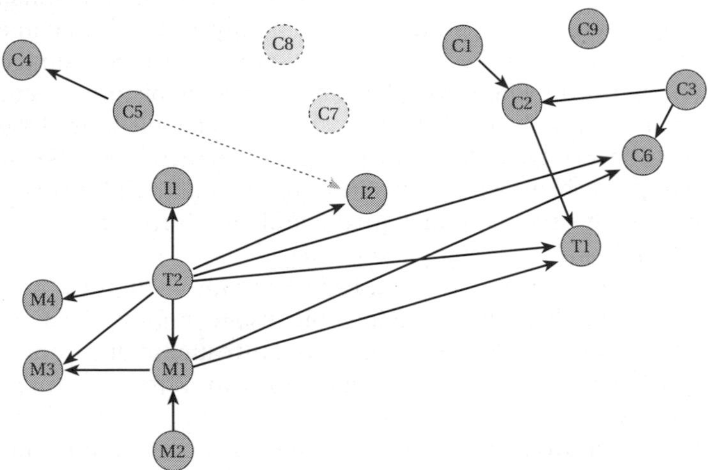

Fig. 5.7. Samsung's regional production network in Asia

Note: The two FDI projects in Suzhou (C7 + C8) have not been realized, but Samsung bought 33 hectares of Suzhou industrial complex and plans to invest more than US$500 million for integrated electronics products from components to end-products (*Business Times*, 19 Sept. 1994).

Source: Dodgson and Kim (1997).

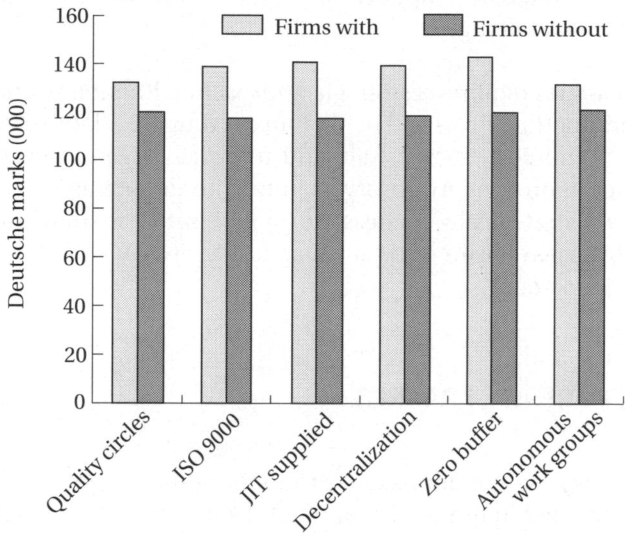

Fig. 5.8. Productivity and the use of modern manufacturing concepts
Note: Value-added per employee; $n = 1{,}305$ capital equipment producers.
Source: Dreher (1996).

strategy of lean thinking has not. In their view, it is the combination of both the ingredients and the overall recipe that produces dramatic results. At the earth-moving equipment company Caterpillar, for example, it used to take 6,000 workers twenty-five days to get a big back-loader through the plant. In the late 1990s it only took 3,000 workers five days, and inventories had been cut by 60 per cent. Similarly it used to take up to ten years to design a new machine, but that has been decreased to twenty-seven months (*The Economist*, 20 June 1998). It follows that, if firms do not take a broader, longer-term view of their operations and production investments, they are unlikely to perceive and receive the value from their investments.

There are a number of other reasons a strategic approach is important for production and operations.

- Production facilities, plant, and equipment generally account for a large proportion of company investments and activities, and it is therefore essential that the decisions companies make about the markets they wish to operate in is informed by production capacity. It follows that there is a need to think of both future and present needs.
- Non-price factors, such as delivery times, reliability, and quality, are increasingly important drivers of competitiveness.
- Operations and production competencies have become a source of competitive advantage through the ability to make and assemble products quicker and cheaper than competitors.

- Massive technological opportunity can be derived from new, flexible production technologies.

Each of these issues requires a strategic approach. Without a strategy for operations and production, firms will be unable to compete effectively and will fail to introduce technology successfully and to obtain a good return on investment. In the opinion of many observers, few firms do manage their operations and production strategically. As Bessant (1991: 1) put it, 'manufacturing history is littered with cases of firms or even whole industries which failed to recognize challenges and to adapt'.

MANUFACTURING STRATEGY

What is a strategy for operations and production? As most of the discussion of this issue is couched in terms of 'manufacturing', this tradition will be continued here. According to Hayes and Wheelwright (1984), there are two important roles manufacturing can offer the strategic strengths of a company. First, manufacturing processes may give the business a distinctive advantage in the marketplace. Pilkington's float-glass process, for example, provided a new production technique that was considerably cheaper and more efficient than that used by competitors. Secondly, manufacturing can provide coordinated support for the essential ways products win orders in the marketplace at a level that is better than its competitors can offer. Manufacturing is, therefore, directly linked to the decisions companies make about which markets they will operate in, and is thus a strategic issue. Manufacturing strategy aims to mobilize manufacturing capacity to gain competitive advantage.

Based on the work of Wheelwright and colleagues, it is possible to consider two broad perspectives, or approaches, to manufacturing investment: those that are 'static' and 'dynamic' (see Box 5.5). Both of these are 'ideal types' on a continuum of actual practice. In the static view of technology, manufacturing uses dated equipment, and when new investments are made they tend to be *ad hoc* (without consideration of the development of the company's skills base or organizational competencies), and are rarely introduced without difficulty. Dynamic approaches, on the other hand, take an evolutionary approach, with investments being made to build strategic benefits. In the dynamic approach, manufacturing investments are made not only to reduce costs, but also to enhance products (Wheelwright 1988; Clark and Wheelwright 1993; and see Hayes *et al.*'s (1988) discussion of 'levels of control').

Hayes and Pisano (1994) take a similar approach to manufacturing strategy. They argue that long-term success requires a firm continually to seek new ways of differentiating itself. Firms that transform their manufacturing organizations into sources of competitive advantage use the various improvement techniques such as TQM and JIT to attain the broader aim of selecting and

Box 5.5. ***Ideal perspectives on manufacturing***

Static perspective	Dynamic perspective
Is typified by command and control over the workforce	Holds the view that the problem of production will never be solved
Views product development as the creative task, and manufacturing purely as operational	Invests in the problem solving skills of the workforce and uses those skills effectively
Has few linkages between R & D and manufacturing	Emphasizes teamwork, both within and between domains
Has little development of new manufacturing processes; automation is primarily a means of reducing costs, of substituting capital for labour	Encourages workers to have a major input in how their work can be carried out—for example, through the use of quality circles
Adopts a hierarchical or vertical view of manufacturing that continually subdivides and specializes the function	Aims to build up internal technological capabilities that can improve incremental innovation, allow shorter development cycles, and improve timing, quality, and cost

developing unique operating competencies. The constituent elements of good manufacturing practice, such as TQM and JIT, should be seen not as ends in themselves, but in terms of the competencies they require and create. And it is these competencies that customers value and competitors should find hard to duplicate. Hayes and Pisano (1994) argue that, in a turbulent environment, the goal of strategy should be strategic flexibility, and the role of manufacturing is to assist this flexibility.

The key elements of manufacturing strategy

According to Bessant (1991), manufacturing strategy can be separated into structural and infrastructural elements. *Structural* elements include:

- make or buy decisions (does the firm manufacture or outsource)?
- number, location, size, quality, and timing of facilities;
- type and capacity of equipment;
- range and extent of support services (e.g. hardware and software maintenance).

Infrastructural elements include:

- systems for controlling production
  - managing quality—TQM
  - introducing new products
  - maintenance
  - information flow management—MRP, JIT
- human resource management
  - quality, age, experience of workforce
  - work organization and coordination
  - communications
  - command and control
- culture—shared beliefs across the firm.

Traditionally, emphasis has been on structural elements—that is, obtaining economies of scale. In the late 1990s, based on Japanese experience, there was greater focus on infrastructure, because the infrastructural issues provide the unique, not easily replicable capabilities that provide strategic advantages.

Bessant (1991) develops a process for manufacturing strategy that includes the following steps:

- *Identify targets for competing* (where do we want to get to?)
  - Understanding basis of competitiveness, separating different product/market families, and examining price/non-price factors.
  - Ascertain the performance required by market
  - Benchmark performance against competitors
- *Audit current manufacturing performance*
  - Examine structure and infrastructure
  - Assess current state of product life cycle
- *Explore options for innovation*
  - Which will enhance strength, overcome weaknesses?
  - What are the benefits of incremental and radical innovations?
  - Understand relative merits and advantages of organizational and technological innovations.
- *Forecasting*
- *Implementation*
  - This is a major source of problems
  - Managing change is the key issue
- *Review*
  - See strategy formulation as a learning process

A brief note on the product life cycle and the manufacturing process used

Products are often described as going through various stages—introduction, growth, maturity, saturation, and decline—in a 'product life cycle'. In

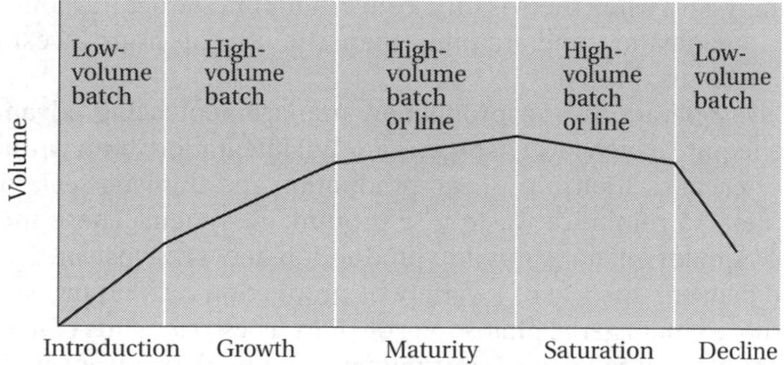

Fig. 5.9. The product life cycle and process choice
Source: Brown (1996).

considering the role the product life cycle has on the manufacturing process used, and the way work is organized around it, it should be remembered that it does not apply to all products. All products do not go through all stages, and some argue that there are product cycles that operate in reverse (Barras 1986; Baba 1989). If the use of the concept of a product life cycle is appropriate, then it is possible to consider the differing requirements of different stages in the product's life (Fig. 5.9). In the early stages of a product life cycle, the production requirement is for flexibility to respond to changing market needs until such time as the product becomes established. There may be many design changes during this period, so there are advantages in production being co-located with R & D. As batches are small and variable, it helps to have a highly skilled workforce operating highly flexible technology at this stage. As the product matures, volume production and cost reduction become a greater concern. Greater automation and lower skilled employees may be used. In later, declining, stages it may be necessary to address particular small market niches, and again the emphasis will be on flexibility.

CONCLUSIONS AND SUMMARY

Operations and production are an important, but often underemphasized, domain. The management of operations and production is critical to firms' competitiveness. If investment in it is undertaken in a strategic manner, it can be an important source of competitive advantage. There are broad international differences in the levels of investment in production technology, and in the way that technology is used. These differences are explained by differences in attitudes towards engineers in industry, investment appraisal techniques, and approaches to automation. When CIM is seen primarily as a labour-saving opportunity, and when there is little concern for infrastructural concerns, such as work organization and training, then the returns from investments are unlikely to be maximized.

Effective operations and production management using advanced technologies require a high level of organizational integration, both within the firm and in external relationships. Lean production and thinking seek methods of organization to minimize waste in a production system. These include new forms of supplier relationships and production networks, just-in-time production and delivery, the use of a variety of production techniques, such as ERP, and the broad managerial philosophy of TQM. These elements of lean thinking can be introduced piecemeal, but the best results derive from their use in a holistic, strategic form, with practices encompassing all the activities of the firm.

Manufacturing strategies can be developed within a continuum ranging from the dynamic to the static. There are advantages for manufacturing efficiencies and effectiveness if the former perspective is adopted and when there is concern to address infrastructural elements. Similar to the issues facing the management of R & D and new product development, the key issues for products and operations are organizational and involve a focus on human resource management and training.

CHAPTER

6

Technology Strategy

WHY IS TECHNOLOGICAL INNOVATION A STRATEGIC MANAGEMENT ISSUE?

Since the late 1980s, an increasing amount of research and analysis has shown why technology should be a strategic issue (Teece 1987; Dodgson 1989; Pavitt 1990; Loveridge and Pitt 1990; Granstrand *et al.* 1992; Dussauge *et al.* 1993; Coombs 1994; Goodman and Lawless 1994; Burgelman *et al.* 1996). As we have seen in preceding chapters, technological innovation is a strategic issue for the following reasons.

- The development and use of technology are a key source of competitive advantage.
- The complex, uncertain, and expensive processes of R & D, developing new products, and production and operations innovations will result in piecemeal, short-term, and potentially disruptive outcomes unless they are guided by a strategy that builds synergies and grows expertise cumulatively.
- The globalization of technology and markets requires companies to take a strategic approach to their technological investments.
- The organizational structures that firms adopt to encourage technological development—for example, in the organization of R & D—should follow the strategy pursued by the firm.

Two additional reasons for strategic management are, first, that existing company strategies that do not integrate technological issues tend to fail, and, secondly, that unless firms can articulate their long-term, strategic aims for technology, they will be unable to communicate with and benefit from public-sector science and technology policies in areas such as basic science, and will fail to build long-term technological collaborations with partners (see Chapters 7 and 9).

WHAT IS TECHNOLOGY STRATEGY?

In common with the literature on corporate strategy in general, there are broadly differing views of what technology strategy is and how easy it is to define, utilize, and alter. For technology-based firms—that is, all firms that rely on technology for their ability to compete—*technology strategy comprises the definition, development, and use of those technological competencies that constitute their competitive advantage.*

In determining what is 'strategic' in technology, the analogy can be made with military tactics and strategy. In military parlance, 'tactical' refers to the means by which battles are won. 'Strategic' refers to how wars are won: whether, where, and when to fight battles; preparation for war by understanding the nature of external threats and opportunities; and ensuring that sufficient resources are collected and organized in order to succeed. For technology, tactical issues include how firms innovate new products and production processes. Strategic matters include choosing and developing the competencies that shape firms' opportunities for innovation and continuing competitiveness. In this schema, many of the issues of managing R & D, NPD, and operations and production discussed in previous chapters are tactical. They become strategic when investment decisions are made in these areas with the intent of extending technological competencies, and when major organizational issues are addressed (through, for example, restructuring to improve inter-functional integration) with the aim of linking product lines more closely with competencies.

For firms relying on technological innovation, successful firms look beyond their individual product lines and build a strategy around core knowledge and competencies, and these provide them with a sustainable and flexible strategic focus (Quinn 1992; Utterback 1994). Technology strategies involve identifying the key technologies that underpin the firm's present and future value-creating activities and ensuring that they are improved, supplemented, and effectively introduced and used. Identifying core technological competencies, however, is much easier to do *ex-post* than *ex-ante*. Many firms are adept at managing some of the ingredients of the strategy recipe, and are comfortable with the concepts of technology planning and assessment, for example. But the most difficult management task, and perhaps the most important challenge of strategy, lies in combining these ingredients with strategic direction and within appropriate organizational structures. Experience shows that, while many firms are aware of the value of utilizing these elements, few are capable of ensuring that activities fit together in a coherent manner with their strategy and structure (Miles and Snow 1994). The pump firm described in Chapter 1 might be excellent at managing R & D, CAD, and quality, but it will not find sustainable competitive advantage until it directs these coherently and cogently into competencies that clearly position it amongst its competitors.

A number of studies have developed typologies of technology and innovation strategy (Goodman and Lawless 1994; Freeman and Soete 1997). Two of the most useful of these are shown in Box 6.1. An 'offensive' strategy involves technical and market leadership with a strong research orientation; 'defensive' strategies focus on defending existing technologies, and incremental improvement; 'imitative' and 'dependent' strategies involve following technological leaders, and focus on production capabilities; 'traditional' strategies involve minimal investments outside the area of production capabilities; and 'opportunist' strategies involve considerable investments in technology search and protection. The Goodman and Lawless typology also maps differences in the market and product situations confronting firms, and approaches to various investments in research, production, and search activities.

All serious studies of technology strategy point to the immense complexity of determining different categories of technology strategy. Few of the existing typologies, for example, include mutually exclusive categories. Analytical approaches to technology strategy mirror the contemporary debate in the corporate strategy field as a whole. This debate revolves around the question of whether strategies should be based primarily on assessment of opportunities and threats in external operating environments, or on the identification and subsequent extension of internal resources, sometimes in the form of core competencies (compare, for example, Porter 1985 with R. Grant 1991, and Prahalad and Hamel 1994).

The difficulties of managing technological innovation strategically, and of assessing external changes and internal resources in order to behave strategically, is shown in Christensen's (1997) discussion of the 'innovator's dilemma',

Box 6.1. ***Typologies of technology strategy***

Goodman and Lawless (1994)	Freeman and Soete (1997)
Technological commodity search	Defensive
Pre-emption	Imitative
Productive efficiency	Offensive
Producer preference	Opportunist
Production flexibility	Traditional
Customer preference	Dependent
Product pioneer/product leader/ product follower	
Vertical integration	
Complementary technology	

which causes the circumstances 'when new technologies cause great firms to fail'. He describes how well-managed firms (those that listen to their customers, invest aggressively in new technologies that provide customers more and better products of the sort they want, carefully study market trends, and systematically allocate investment capital to innovations that promise the best returns) consistently lose their positions of leadership in a wide range of industries. Thus, in the development of the computer disk-drive industry, the leaders in one generation of disk drives (14″, 8″, 5.25″, etc.) were, in every case, displaced in the next.

Christensen analyses three elements of a 'failure framework' that help to explain the seeming paradox of well-managed firms failing. First, he explores the distinction between *sustaining* and *disruptive* technologies. Most new technologies, he argues, are sustaining, and improve product performance. Disruptive technologies, on the other hand, may initially underperform compared with existing products, but may have advantages that appeal to a new set of customers. Eventually, the disruptive technology replaces existing technology more widely, as the transistor replaced the vacuum tube, and the PC the mainframe. Christensen found that sustaining projects addressing the needs of the firm's most powerful customers almost always pre-empted resources from disruptive technologies with small markets and poorly defined customer needs. Essentially, successful companies eventually fail because they find it difficult to invest in disruptive technologies.

A second aspect of failure explored by Christensen is the situation when technological progress in successful firms outstrips market needs. The third aspect he analyses is how investment patterns in successful companies are driven by existing customers and financial structures.

Christensen provides some practical answers for dealing with the innovator's dilemma and makes some important observations about the role of learning. Most managers, he argues, learn about managing *sustaining* technology, when markets are targeted, and when the planned approach to evaluating, developing, and marketing innovations is essential. However, the capabilities, cultures, and practices that underlie this form of management are valuable only in certain conditions. In contrast, Christensen (1997: 147) argues, 'not only are the market applications for disruptive technologies unknown at the time of their development, they are unknowable. The strategies and plans that managers formulate for confronting disruptive technological change, therefore, should be plans for learning and discovery rather than plans for execution.'

This assertion has some resonance with Granstrand and Sjolander's (1990) research on 'multi-technology corporations', which argues that, as firms face increasing numbers of technological transitions (moving from one generation of technology to another), they move up a 'managerial learning curve'. In this way, the more transitions they experience, the better they become at managing them, and hence improve their company's competitive strengths. Similarly, Tushman and O'Reilly (1997) contend that long-term organizational success depends on streams of systematically different innovations over time. These streams prevent organizational inertia, and require the development of what they call 'ambidextrous organizations'—that is, organizations that celebrate stability and incremental change as well as simultaneously encouraging experimentation and discontinous change. This capacity to learn and develop the organizational techniques to deal with ranges of technology explains why, historically, technological leaders manage to maintain some elements of their leadership during periods of rapid change (Cantwell and Anderson 1996).

Assessing technological opportunities and threats in the external environment is notoriously difficult. Even the most knowledgeable people get their predictions about technological development horribly wrong (Box 6.2). The

Box 6.2. ***The dangers of predicting technology development***

'Heavier-than-air flying machines are impossible.'

(Lord Kelvin, President, Royal Society, 1895)

'Airplanes are interesting toys but of no military value.'

(Marechal Ferdinand Foch, Professor of Strategy, École Supérieure de Guerre)

'This "telephone" has too many shortcomings to be seriously considered as a means of communication. The device is inherently of no value to us.'

(Western Union internal memo, 1876)

'Everything that can be invented has been invented.'

(Charles H. Duell, Commissioner, US Office of Patents, 1899)

'Computers in the future may weigh no more than 1.5 tons.'

(*Popular Mechanics*, 1949)

'I think there is a world market for maybe five computers.'

(Thomas Watson, Chairman of IBM, 1943)

'But what . . . is it good for?'

(Engineer at the Advanced Computing Systems Division of IBM, 1968, commenting on the microchip)

'There is no reason anyone would want a computer in their home.'

(Ken Olson, President, Chairman and founder of Digital Equipment Corp, 1977)

'640K [of RAM] ought to be enough for anybody.'

(Bill Gates, Chairman of Microsoft, 1981)

Source: Innovation Management Network: http://mint.mcmaster.ca.

uncertainty and unpredictability of technological change make accurate assessments of its development the result more of serendipity rather than of prescience. According to Pavitt (1994: 358), 'innovative activities have remained highly uncertain in their commercial outcome . . . In addition, both practitioners and theorists still have great difficulties predicting the rate and direction of radical technical change.' Given the immense difficulties in predicting the nature of technological opportunities and threats, the primary, but certainly not exclusive, focus of technology strategy should be on the development of internal competencies.

WHAT ARE TECHNOLOGICAL COMPETENCIES?

Competencies consist of two elements: the *resources* currently available to a firm, and the *innovative capabilities* the firm possesses to define and change those resources. Resources comprise all those elements of the firm that enable it to function—people, equipment, organizational routines, finance—and provide the basis for future developments. They are, therefore, a *static* concept, existing in one particular point in time. Innovative capabilities include a range of activities—searching, acquiring, implementing, integrating, coordinating, and learning—and are *dynamic* in nature, allowing firms to transform themselves by utilizing the options created by the resource base.

The distinction between resources and innovative capabilities has a relatively long history. Edith Penrose's (1959) classic study, *The Theory of the Growth of the Firm*, carefully distinguishes 'resources' from 'services' (innovative capabilities). She conceives of resources as providing the basis for production, and of services as the means of activating the potential that resources provide. This distinction is continued in recent work on strategy, including resource-based theory of the firm (R. Grant 1991), and dynamic capabilities theory (Teece and Pisano 1994). Some of the major elements of resources and innovative capabilities are described in Box 6.3. There is a great deal of terminological confusion in this area between the use of 'capabilities' and 'competencies', and the prefix 'core' is often used. Here the term 'competencies' is used as the central element defining the firm's ability to compete, and 'technological competencies' refers to those competencies that have a technological basis. Differential competencies allow a firm to gain benefits over competitors. Competencies have strategic potential when they are

- *valuable*—exploit opportunities and/or neutralize threats in a firm's environment;
- *rare*—the number of firms that possess them is less than that needed to generate perfect competition in an industry;
- *imperfectly imitable*—because of their complexity, or the uniqueness of the conditions under which they were acquired;

Box 6.3. ***Conceptualizing the resources and innovative capabilities that define technological competencies***

Resources	Innovative capabilities
• *Resident knowledge and skill* ∘ amongst managers and other personnel • *Company routines* ∘ patterns and plans of behaviour and organization • *Technology base* ∘ plant and equipment ∘ R & D facilities • *External networks* ∘ user/supplier ∘ horizontal (joint ventures etc) ∘ academic • *Financial resources*	• *Forecasting and assessing* • *Searching and selecting* ∘ Information, technology and markets • *Acquiring and protecting* ∘ new resources and options • *Implementing* ∘ change and new strategies • *Aligning* ∘ technology plans to technology audits ∘ technology and business strategy • *Integrating* ∘ different functions and divisions ∘ external and internal inputs • *Coordinating* ∘ imaginative and entrepreneurial combining of resources to create comparative advantages • *Learning*

Source: Dodgson and Bessant (1996).

- *have no strategically equivalent substitutes*—no alternative ways of achieving the same results (Ciborra and Andreu 1998).

The distinction between resources and innovative capabilities helps explain long-term business and industrial dynamics. Firms with similar resource bases often do not fare similarly in the long term. It is the firms with existing resources and the innovative capabilities to change those resources to deal with changing opportunities that succeed. Resource-rich firms, like IBM in the 1980s, could not compete with the new entrants because they lacked appropriate innovative capabilities.

By providing the focus of technology strategy, technological competencies are the basis by which targets are set for the development, acquisition, diversification, and disposal of different resources and innovative capabilities. They provide a focal point for decisions about whether to lead in the development of a technology, or follow others, and whether to collaborate, purchase, or develop it internally.

Resources

Firms do not possess unlimited opportunities in the development of their technology strategies. Technology strategies, and the options firms have, are constrained by the resources they possess. These resources not only include technological assets, like R & D facilities and factories, financial assets, and human resources; they also include the networks to which the firm belongs, and the 'routines' they follow in their organizational behaviours and practices (which are often difficult to change).

Thus a large chemical firm cannot easily become a car-manufacturer nor can a small furniture-maker readily produce steel. As theoreticians such as Dosi (1988) and others have pointed out, what a firm has done in the past determines what it can do in the future. The empirical research of Patel and Pavitt (1998) shows that firms' main products strongly influence the directions for future technological change. Change is always possible, of course, and many companies successfully diversify technologically, but it does not happen easily or quickly. This point also applies to the routines firms use. Highly formalized organizations cannot overnight assume the characteristics of small, entrepreneurial firms.

Different types of technology firms face different constraints, opportunities, and challenges. Pavitt (1994) uses a typology of technology businesses to distinguish the differences they experience in sourcing technology and the main tasks they face in their technology strategies (Box 6.4). Pavitt (1998) extends his analysis to consider the strategic options best fitted to firms with different types of technology base. He suggests that firms with low technological opportunity are more likely to be compatible with an 'administrative' rather than an 'entrepreneurial' management style. When these firms have high costs of investment, such as in the steel industry, there is likely to be more centralization compared to industries where there are low costs of investments and conglomerates are a more suitable form. Firms with high technological opportunity and high costs of product and process development, such as drugs and automobiles, are likely to be best suited to a strong entrepreneurial function at the corporate level. Firms with high technological opportunity, but low costs of product development, such as in consumer electronics and the 3M corporation, will be best served by more decentralized entrepreneurship (Fig. 6.1).

It is important to note that firms may move between the quadrants shown in Fig. 6.1. For example, GEC in the UK and ITT in the United States moved from the entrepreneurial to the administrative form of corporate control as they reduced their commitments to R & D. Similarly, mainframe computer manufacturers moved from circumstances of very high costs of NPD in the 1970s to lower costs, leading to a more divisional form of organization.

Innovative capabilities

Competencies involve all those activities that enable firms to change their technological resources—that is, their innovative capabilities. These include

Box 6.4. *Types of technology firm*

	Supplier-dominated	Scale-intensive	Information-intensive	Science-based	Specialized suppliers
Typical core sectors	Agriculture Services Traditional manufacturing	Bulk materials Automobiles Civil engineering	Finance Retailing Publishing Travel	Electronics Chemicals Drugs	Machinery Instruments Software
Main sources of technology	Suppliers Learning from production	Production engineering Learning from production Design offices Specialized suppliers	Software and systems departments Specialized suppliers	R & D Basic research	Design Advanced users
Main tasks of technology strategy	Use technology from elsewhere to strengthen other competitive advantages	Incremental integration of changes in complex systems Diffusion of best design and production practice	Design and operation of complex information processing systems Development of related products	Exploit basic science Development of related products Obtain complementary assets Redraw divisional boundaries	Monitor advanced user needs Integrate new technology incrementally

Source: Pavitt (1994).

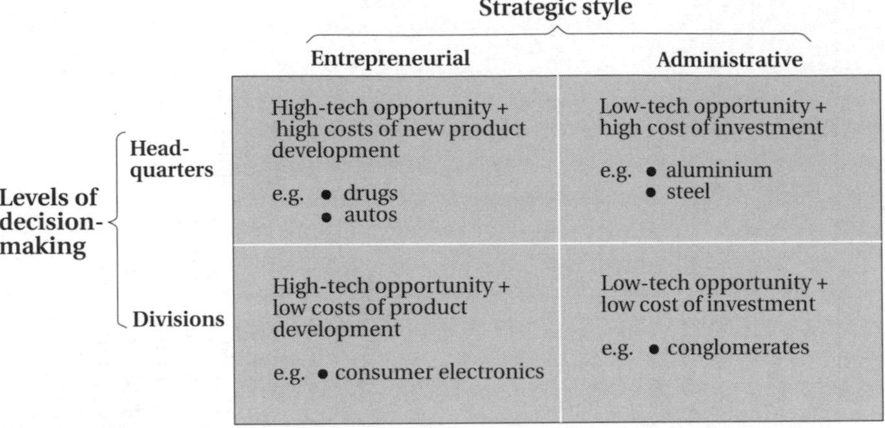

Fig. 6.1. Technology and corporate control
Source: Pavitt (1998).

the ways firms *forecast* and *assess* changes occurring in their technologies and markets.

Forecasting and assessing

As we have seen, identifying future technological development is a necessarily speculative exercise, but it is valuable none the less. Forward-thinking firms welcome any information, guidance, or advice on likely future developments or scenarios in their areas of science and technology, and on the trajectories their technology is likely to follow (Coombs 1994). Firms can use methods such as the Delphi technique of attempting to build a consensus of opinion about future developments amongst leading researchers and thinkers. Other firms use scientific advisory boards.

NEC's approach to technology forecasting involves the identification of over thirty core technologies followed by attempts to predict developments in their underlying basic technologies. The core technologies provide the basis for future research and applications. For example, NEC has identified pattern recognition as a core technology (Fig. 6.2). It then attempts to predict changes in the underlying basic technologies of pattern recognition, and any new product opportunities that might emerge from those changes.

Assessment of the external context in which firms operate includes consideration of issues such as the systems in which firms operate, including the national innovation systems, and the networks in which firms belong, and the ways in which these are affected by globalization (see Chapter 2). Industry has played an important role in the series of government-sponsored 'technology-forecasting' exercises that have taken place in an increasing number of nations

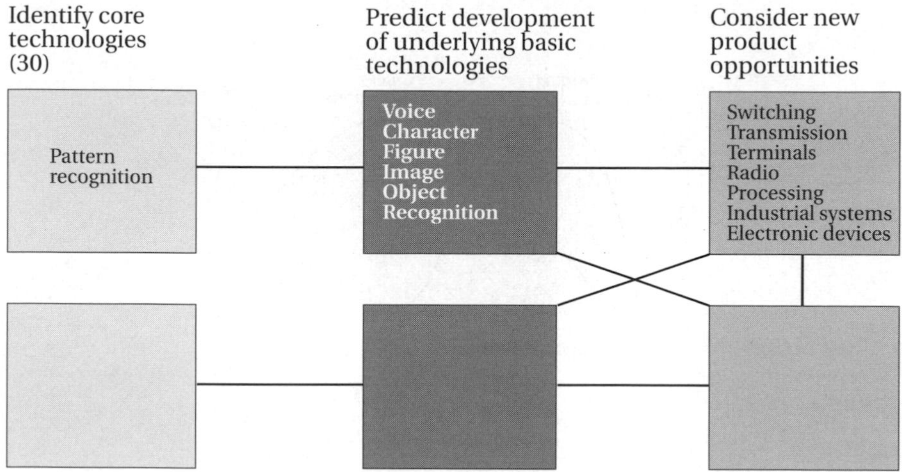

Fig. 6.2. Technology monitoring and forecasting in NEC
Source: based on Irvine (1988).

(Martin and Johnston 1998). These exercises involve industry, government, and academia joining together in seeking a convergence of opinions about the importance of key future science and technologies. Industrial participants have found the process valuable in enabling them better to understand the potential opportunities and challenges confronting them in science and technology.

The process of assessment is assisted by consideration of life cycles for the industry, for innovation, and for the so-called S-curve. These are illustrated in Fig. 6.3. As we saw in Chapter 5, these curves do not apply to all innovations and industries, but they are useful analytical tools which help a firm to consider its relative position before making investment decisions.

Searching and selecting

Innovative capabilities include the methods by which firms *search* for and *select* technologies that will provide the future basis of market competitiveness. Search is assisted by the conduct of basic research in companies, which, as we saw in Chapter 3, sends messages to external researchers that they are interested in a particular science or technology. As a result there may be potential receptors for the research being conducted. Patent searches provide a useful method for assessing a technology and how strong competitors are in a particular area. Bibliometric techniques—examining publications in the scientific literature, and the extent to which they are cited—provides another method of searching for new opportunities. It is not only academics who publish in the

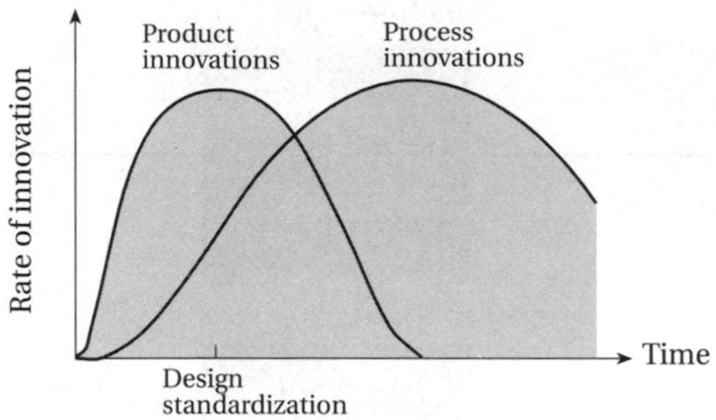

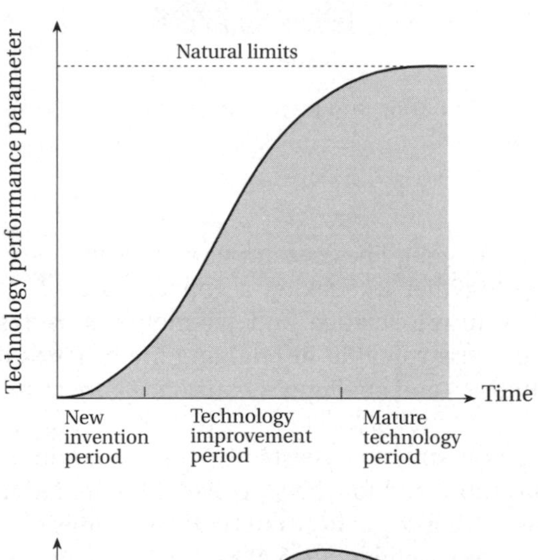

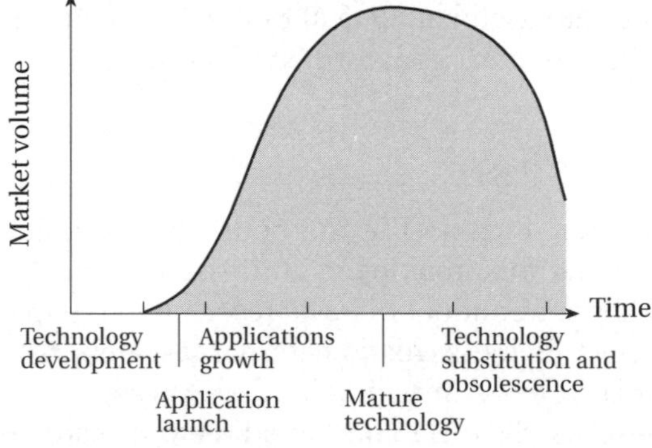

Fig. 6.3. Product and industry life cycles
Source: Betz (1996).

science literature; firms also publish extensively (Hicks 1995). Other information is obtained from a variety of sources, ranging from attending conferences and exhibitions to industrial espionage.

The selection of new technologies can be the responsibility of central R & D, decentralized, business-unit R & D, or one of the committees designed to oversee corporate R & D projects (discussed in Chapter 3). For firms selecting future technological investments, it is worth referring to studies that show

- the importance of cumulative know-how, and the value of building on existing competencies (Prahalad and Hamel 1994);
- that complete technological diversification is extremely difficult and usually unsuccessful, whereas incremental diversification is the least risky (Roberts 1991).

Honda provides an example of the way firms can think about competencies. It produces a variety of products, but its core expertise lies in the design and manufacture of engines and powertrains, and it has developed its product range around these competencies (Fig. 6.4). Once core competencies have been determined, it is possible to consider their technological basis (Fig. 6.5). Using this form of analysis, it can be seen that Honda's technology strategy involves the supplementation and improvement of those technologies that contribute to its core competencies.

Sumitomo Electric Industries provides an example of related diversification. According to Kenney and Florida (1994: 319),

> the company's original core business in the late nineteenth century was copper mining and smelting. During the early part of the twentieth century Sumitomo Electric used its skills in smelting to move its business focus progressively into copper wire manufacture. In the immediate postwar era the company drew upon its technological base in metals and wire to move into special steel wires, in wire coatings to move into rubber and plastic products, and in electronics to move into electronic materials and antenna systems. By the 1960s, the company moved into progressively more complex systems

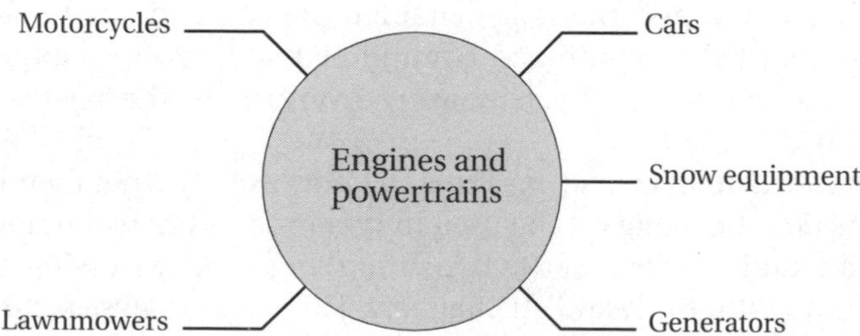

Fig. 6.4. Honda's core competencies

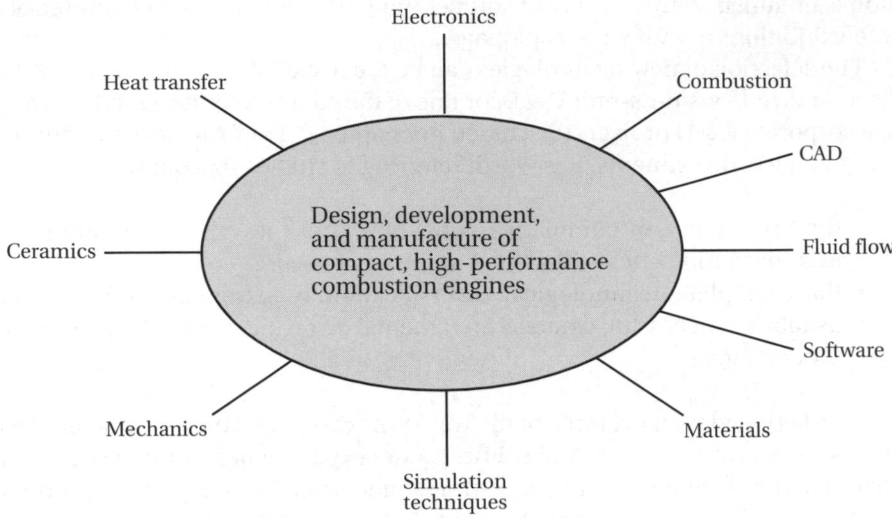

Fig. 6.5. The technological basis of Honda's core competencies
Source: Granstrand *et al.* (1992).

technologies such as integrated electronics systems and disc brakes. In the 1970s and 1980s, Sumitomo used its built up technological competencies to move into high-technology electronic systems (e.g. workstations) and automotive electrical systems. At each stage in its development, Sumitomo leveraged internal technological capabilities developed through R & D to underpin diversification and growth by entering new fields.

3M provides another example of a firm that historically has developed around core technologies. Its original core technologies, developed in the 1920s and 1930s, of abrasives, adhesives, and coating-bonding have been continually developed and added to by four new core technologies: software, instrumentation, imaging, and 'non-woven' (for example, fabric and film). As Quinn (1992) argued, these core technologies (some of which were acquired rather than developed internally) enabled the company to develop a range of products that led to continued commercial success. In contrast, when the company attempted to diversify away from them, the results were not so successful.

It is important, of course, for firms to know exactly what their technological strengths are. Tools that can be used in this process are technology audits (see Goodman and Lawless 1994), benchmarking, and assessing technological strengths quantitatively and qualitatively. These can analyse factors such as the comparative quality and productivity of researchers in similar fields (publications and patents per employee), the range and depth of patent ownership, and manufacturing productivity.

Acquiring and protecting

Innovative capabilities also include the means by which firms *acquire* new technological resources through R & D, licensing, and direct purchase. David Teece's (1986) analysis found that every innovation is accompanied by a number of 'complementary assets' required for its commercialization, such as competitive manufacturing capabilities, sales, distribution and services, and complementary technologies (Fig. 6.6). In this form of analysis,[1] there are three major types of governance structures that can link the sources of R & D with the assets necessary for their commercialization—arm's-length transactions, vertical integration, and collaboration—each of which has advantages and disadvantages.

Arm's-length transactions assume minimal relationships between suppliers and customers and can be considered akin to spot-trading. *Vertical integration* involves the conduct of production and distribution activities within a single firm. The *collaboration* form of governance structure manages to merge some of the advantages, and overcome some of the disadvantages of both

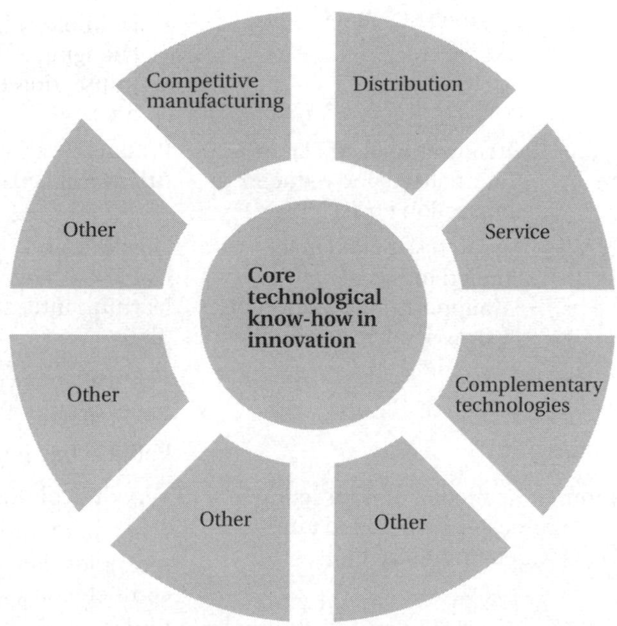

Fig. 6.6. Representative complementary assets needed to commercialize an innovation

[1] The group of scholars undertaking this analysis can be referred to as the 'Berkeley School' and includes David Teece, David Mowery, and Oliver Williamson (see Williamson 1975, 1985; Teece 1986, 1987; Mowery 1987, 1988).

Box 6.5. ***Governance structures linking R & D with complementary assets***

Governance structure	Advantages	Disadvantages
Arm's-length transactions	Useful when technology is • codified (e.g. can be expressed in tangible media • discrete (e.g. separable) • appropriable (e.g. protectable by IPR or other methods) • non-complex • mature	Cannot control opportunism by other firms Information has to be given to a large number of firms Possibility of reverse engineering Cannot unbundle technology, and ascertaining value is difficult
	Useful when technology has • long product cycles • low transaction costs	Problematic if • patent protection is weak or product is complex • technology is rapidly changing • transaction costs are high
Vertical integration	Organizational efficiency (internalized and reduced transaction costs)	Productive efficiency (no specialisms)
	And in respect of Direct Foreign Investments • Supports exploitation of firm-specific capabilities	High risk and cost, especially with large-scale, learning-intensive plant Political and economic uncertainty Political barriers
Collaboration	Reduction of transaction costs compared to arm's-length transactions	Very difficult to manage Problem of matching technological competencies
	Can account for firm-specific (non-codified) skills	Limited partners (dependency)
	Avoids uncertainty in ascertaining value	Problem of matching company cultures
	Avoids opportunism (limited partners, and know-how loss	

Governance structure	Advantages	Disadvantages
	compensated by financial commitments)	
	Good monitoring of partner's behaviour	
	Technology can be 'unbundled' – it can be selectively transferred	
	Cheaper alternative to merger (covering only limited functions and products)	
	Allows quick access to technology	
	Lower financial risks	
	Lower political risks	
	Better access to foreign markets than exports allow	
	Allows long-term trust to develop (improved products and processes can be dealt with)	
	Allows long-term learning	

arm's-length transactions and vertical integration (these will be examined in greater detail in Chapter 7). As Box 6.5 indicates, the effectiveness of a governance structure is influenced by technology. For example, whether a technology is codifiable and appropriable affects the governance structure's capacity to acquire and protect innovations.

The issues of protecting innovations through IPR and standards will be examined in Chapter 8. As we saw in the case of the IBM PC, IBM's inability to control the IPR in strategically important components led to its eventual difficulties. IPRs are an important strategic concern.

Implementing

Innovative capabilities include the effective utilization and *implementation* of technology. Implementation is much easier when there is broad agreement about the importance of particular technologies and their relevance for the company's future. Reaching such agreement is assisted by the new paradigm of management, and its approach to strategy, outlined in Chapter 2. This approach focuses on gaining consensus and commitment to decisions. Whilst this process may be lengthier than simple top-management fiat, implementation of decisions is faster, as there is usually less opposition to decisions that are fully understood.

Some companies develop 'technology plans'. These can help an organization take an overall perspective of its technological activities. It can establish priorities with business managers for what technical programmes will be pursued, and it can help forecast future developments of technologies and their social and economic impact and influence. A technology plan involves assessment of

- an R & D organization's capabilities;
- the immediate and strategic business needs that an R & D organization must serve;
- the future technical potential of various technologies;
- the leverage that those technologies could provide in business applications (Szakonyi 1990).

Coordinating and integrating

Innovative capabilities include the ability to *coordinate* and *integrate* all the functions of the firm around its technological activities. Organizational and technological integration facilitates flexibility, responsiveness, and growth (Whiston 1991), and is a key element of technology strategy. The strategy pursued affects the structure of R & D and its centralization/decentralization as discussed in Chapter 3. In his study of eighteen leading European and Japanese multinationals, Reger (1997) found that firms used a variety of structural and formal mechanisms to integrate R & D into corporate strategies, including the use of strategic committees, strategic planning departments, and R & D and technology portfolios. Reger (1997: 311) cites ABB as an example of a company that uses a chief technology officer (CTO) as coordination instrument.

> The CTO co-ordinates all corporate-wide research activities via the 'Corporate Research Steering Committee', elaborates on the enterprise's technology strategy together with the Senior Vice-Presidents (SVPs) of the four divisions, and is the overseer of interdisciplinary, cross-enterprise projects. One important instrument used by the ABB is the drawing up of an annual R & D and technology strategy that is closely linked to business strategy at the level of divisions, groups and the concern. This is performed by the Corporate Research Steering Committee which includes the SVPs and the CTO. First, a research and technology review is made and a ranking of current R & D projects is carried out. Then the R & D budget for the following year and a three-year plan are drawn up. Parallel to this process, the Technology Committees of the divisions, each headed by the relevant SVP for research and technology, would draw up a technology strategy and R & D portfolio at the divisional level every year. The CTO functions as an integrator. To support current work, the CTO has a small technology planning team at his disposal at the ABB headquarters.

Alcoa has a technology board comprised of the CEO, senior business leaders in the company, and the CTO. Technology strategies in the company are led by 'Technology Management Review Boards' (TMRBs), which are comprised of senior staff from business units with similar technologies, such as extrusion, chemicals, and smelting. There are a number of cross-TMRBs, which work internationally at spreading good practices in areas such as product design. Technology strategies in the TMRBs, aligned with the business units' needs, are developed and reviewed by the Alcoa Technology Board (Smith *et al.* 1999).

Another means of assisting coordination and integration at a corporate level is the corporate unit for the strategic management of technology. The functions of this type of unit are

- to analyse the structure of the overall technology portfolio;
- to ensure that a technological competence in one business is known to and available to other potential user businesses in the group;
- to identify technical competencies that straddle businesses and to take steps to strengthen them through horizontal organizational links and through small special budgets;
- to consider the overall technology portfolio and inject an appreciation of this portfolio into the broader strategic management processes of the company (Coombs and Richards 1991).

Integration is also required across technologies, and firms use a variety of hybrid instruments (see Chapter 3) to facilitate cross-disciplinary research. Examples include Hitachi's 'core projects and programs' and NEC's 'core technology program' (Reger 1997).

One way of conceptualizing the level of integration required, and the sort of technology strategies that could be pursued, involves distinguishing whether the underlying technology being used, the product or process, and the eventual market are *new* or *existing*. Some examples are shown in Fig. 6.7, which considers a number of strategic archetypes. A new venture strategy may require the integration of new technology, new products and processes, and the creation of a new market in, for example, multimedia. Groupware is a new technology in a new market, using existing product platforms. The production of insulin through biotechnology provides an example of a strategy using a new technology in an existing product to extend an existing market. PCs for use in the home use existing technology in a new product and market. Fast trains, such as the French TGV, use existing technology in a new product to meet existing demand.

Aligning

Innovative capabilities include the ability to align the activities of the firm to unified purposes. This is particularly important in ensuring the alignment of technology and business strategies. A number of simple tools have been developed to help align technology and business strategy. Some of these orient R & D to business needs; others consider business opportunities emerging from R & D.

Three tools for orienting R & D to business needs are discussed here. One tool, shown in Fig. 6.8, involves determining the firm's current competitive

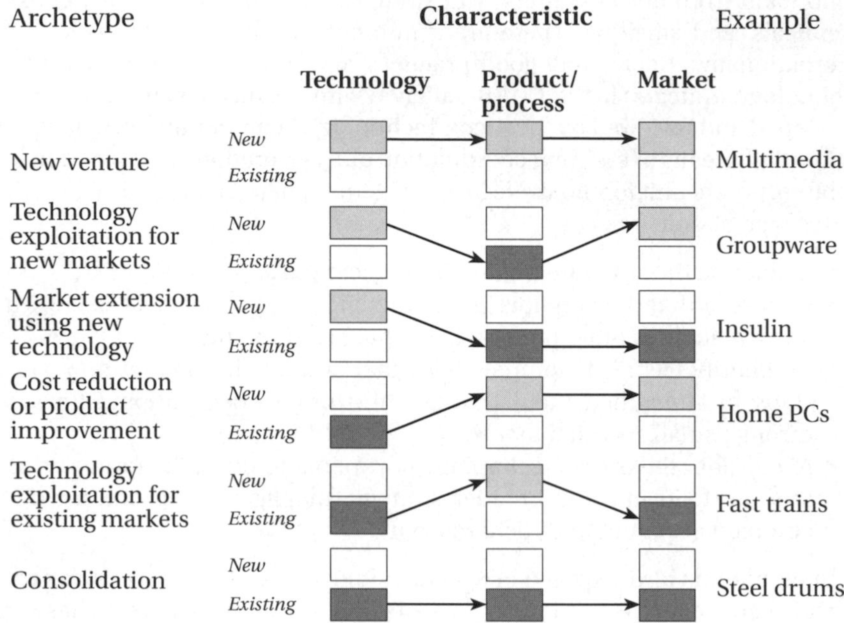

Fig. 6.7. Strategic archetypes
Source: based on EIRMA (1990).

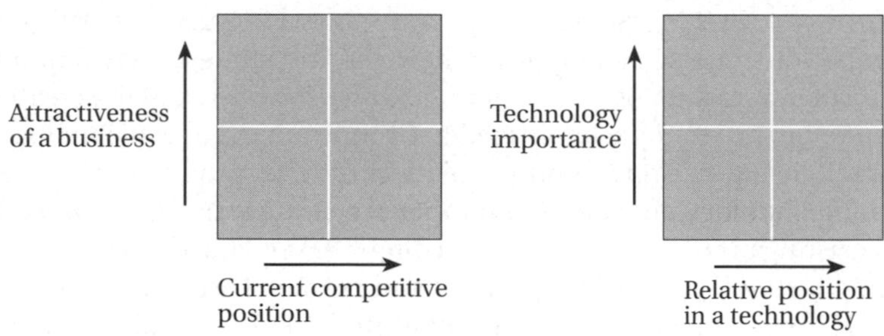

Fig. 6.8. A tool mapping competitive position, business attractiveness, and technology
Source: Harris *et al.* (1993).

position and the attractiveness of a business (likely future profits or growth), and then integrating technological considerations by evaluating the importance of technology for future business and the firm's relative position in that technology. If a firm sees an attractive business opportunity in an important technology, but has a weak relative position in the technology, it needs to invest in the technology to meet its strategic objectives. The tool shown in Fig. 6.9 can be used to map a firm's market-share strategy and product-line needs.

If it wishes to enter new markets with new products, then the focus of the strategy should be R & D. Alternatively, if it wants to maintain or yield market share in existing products, no investment in technology is necessary. The tool shown in Fig. 6.10 maps customer demands from products with the level of technology. If customers demand novelty in high-tech products, the strategic focus should be on R & D. If they demand stability in high-tech products, the focus should be on process engineering. In low-technology products,

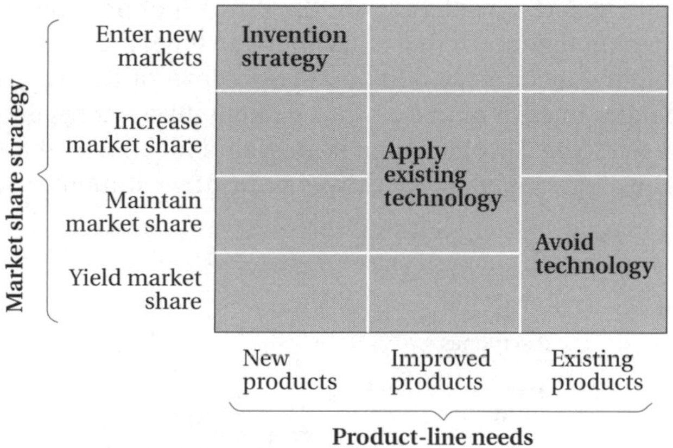

Fig. 6.9. A tool for mapping a firm's market-share strategy and product-line needs
Source: Bitondo and Frohman (1981).

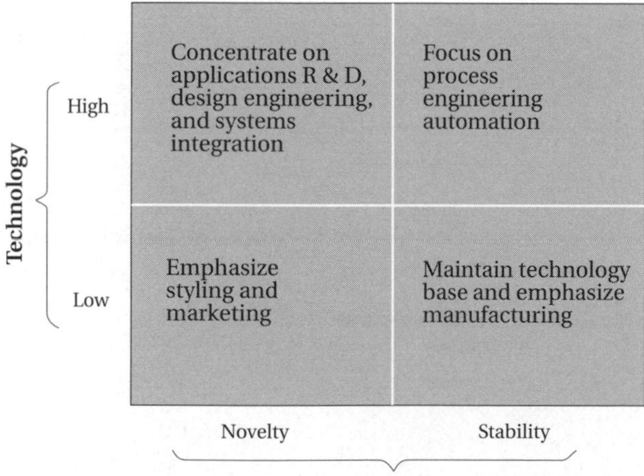

Fig. 6.10. A tool for mapping customer demand and level of technology
Source: Lauglaug (1987).

customers' demand for novelty can be met by styling and marketing. If stability is demanded, manufacturing is again the focus, but less technological investment is required.

A second set of tools, orientated towards allowing R & D to create new business opportunities or advantages, is shown in Figs. 6.11–6.13. The tool shown in Fig. 6.11 involves auditing the technical skills and disciplines existing in a firm, then considering the areas where these can be applied, the particular products and services in which they are used, and the specific market needs being addressed. It then considers new applications for those skills and expertise. The model shown in Fig. 6.12 considers seven dimensions of product acceptability and evaluates technology according to its ability to meet these requirements. It involves considering technology-demand elasticities—or the way markets will respond to changes in each of these—and making investments accordingly. If reliability were felt to be a problem and improvement could increase customer demands at a level greater than investments in other dimensions, then this

Fig. 6.11. Technical audit to discover new applications
Source: Mitchell (1986).

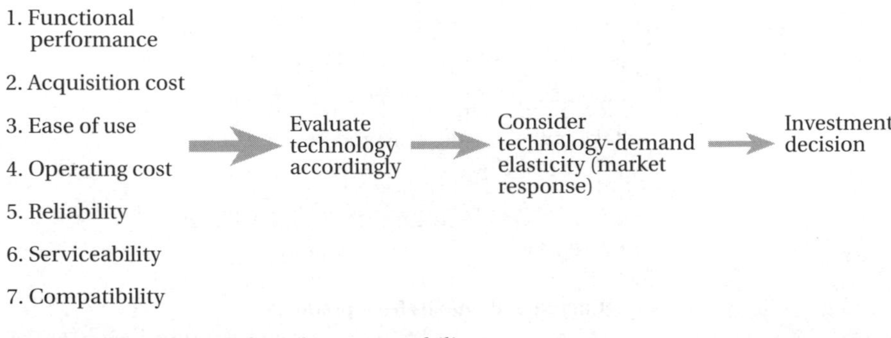

Fig. 6.12. Dimensions of product acceptability
Source: Fusfield (1978).

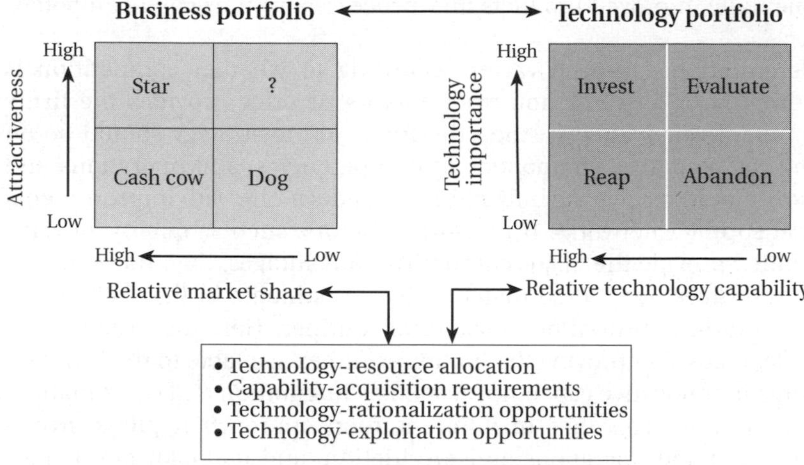

Fig. 6.13. Managing the innovation stream
Sources: Arnold *et al.* (1992).

should be a target for investment. The tool shown in Fig. 6.13 is a sophisticated version of the simple business portfolio analysis, as it integrates the technology-portfolio and technology-strategy options. Thus a 'star' in the business portfolio, and 'invest' in the technology portfolio, imply investments in technological resources. A 'cash cow' and 'reap' portfolio involves strategies for technology exploitation. A 'dog' and 'abandon' imply technology rationalization.

BALANCING INVESTMENTS IN RESOURCES AND INNOVATIVE CAPABILITIES

To develop their technological competencies, firms need to balance their investments in resources and innovative capabilities. These investments should reflect the firm's broad need to *consolidate*, *differentiate*, or *diversify* its technological competencies (there is overlap between these approaches, and companies can pursue elements of each).

A consolidation approach essentially means that companies are contented with the competitive advantages their existing technologies confer, and their technological investments are designed to maintain their current position. Such a strategy will focus on the resources component of company competencies. It will involve a generally low level of strategic investment in R & D, NPD, and operations (that is, investment to maintain the current technology base), and the primary intent of the investment is to ensure a continuing flow of incrementally improved products and processes from existing competencies.

A differentiation approach varies according to whether competition is based primarily on price or non-price factors. If price provides the firm's major competitive advantages, then the focus of the strategy should be on improving the resources component of competencies, and operations and production should receive the bulk of investments along with improvements in external supplier networks. If non-price factors, such as quality, novelty, and flexibility, provide the major competitive advantages, then these investments in resources need to be matched by investments in the skills of the workforce, and in improved organizational routines. Here the primary, but not the sole, focus is improving the broad resources available to the firm; but investments in innovative capabilities, such as integration and coordination, are also important. This type of differentiation approach requires investments in R & D and operations and production, and its major aim is new product development.

A diversification approach focuses primarily, but not solely, on innovative capabilities—those factors that enable the firm to do things differently. When the intent is related diversification, implementing, aligning, integrating, and coordinating activities are central, and the bulk of investments should be in applied R & D. When the intent is the creation of new competencies, then the focus should be on forecasting, searching and selecting, and acquiring and protecting tasks. A substantial focus of R & D should be on the creation of new options, through the conduct of basic R & D and links with the science base.

A simple model to assess where the balance of investments should be made is shown in Fig. 6.14.

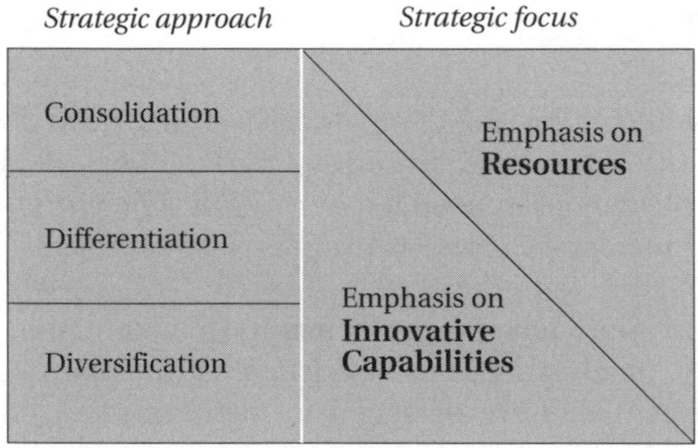

Fig. 6.14. Assessing the balance of investments

A brief note on dynamic-capabilities theory

Dynamic capabilities theory (Teece et al. *1990; Teece and Pisano 1994) is a particularly valuable approach to understanding firm competitiveness and strategic management. It is very helpful in unifying many theoretical approaches and their application to technological innovation. The focus of dynamic-capabilities theory is 'the mechanisms by which firms accumulate and dissipate new skills and capabilities, and the forces that limit the rate and direction of this process'. It emphasizes the changing nature of the environment (the sorts of changes outlined in Chapter 2), and the key role of strategic management in appropriately adapting, integrating, and reconfiguring internal and external organizational skills, resources, and functional competencies in response to those changes.*

Based on traditional theory of the firm, dynamic-capabilities theory argues that competitiveness is derived from the ability to reproduce distinctive organizational competencies over time. It also builds on recent theory of the firm and encompasses concepts such as path dependencies, technological opportunities and timing, transaction costs, asset configuration, and selection environments.

The notion of path dependencies *refers to the way a firm's history helps define and direct future activity.* Technological opportunities *and* timing *are a function of firms' R & D budgets and project selection, developments in the science base, and the way large firms can assist the direction of technological development. A salient concept here is that of 'technological trajectories' (Nelson and Winter 1982). In the circumstances of specific firms, these refer to the way that technology incrementally develops within a firm, and reflect the technology's relationship with endogenous factors (such as the firm's cumulative learning abilities) and exogenous factors (such as market pressures). Timing also emphasizes the importance of chance events, identified by Arthur (1990) as being so important in economic development.*

Firms possess technological, complementary, financial, *and* locational *assets.* Transactions costs *introduce the question of relative cost efficiencies of differing organizational forms.* Selection environments *essentially refer to the degrees of freedom companies have in operations, and include factors such as the extent of competition, capital constraints, and path dependence. Central to dynamic-capabilities theory, and to our analysis of innovative capabilities, is the issue of learning.*

LEARNING AND TECHNOLOGY STRATEGY

The ways firms learn, and use that learning, have become key elements of analyses of corporate strategies (Senge 1990; Howard 1993). The concept of learning is central in Fruin's (1992) analysis of the Japanese enterprise system. 'Technological learning' is argued to be an important element of firms' development and survival (Hayes *et al.* 1988; Dodgson 1991). Learning can be described as the ways firms build, supplement, and organize knowledge around their competencies and within their cultures, and adapt and develop organizational efficiency through improving the use of these competencies. The need to learn is commonly explained in terms of a requirement for *adaptation* and improved *efficiency* in times of change. In turbulent environments learning can be seen as a purposive quest by firms to retain and improve competitiveness, productivity, and innovativeness. The greater the uncertainties facing firms, the greater the need for learning. Learning can be seen to have occurred when organizations perform in changed and better ways, and when competencies are better defined, more appropriate, and effectively implemented. The goals of learning are useful outcomes, which include, at best, improved competitive performance, and, at worst, survival.

Firms need to learn for a number of reasons. They need to learn so as to respond to changes in the external environment—for example, to the rapid and comprehensive changes occurring in technology described in Chapter 2. And they need to learn so as to overcome strategic and organizational tendencies towards introspection and parochialism (Morgan 1986), which can be particularly disadvantageous in turbulent and rapidly changing circumstances.

Learning is a multifaceted and complex concept. It can be *contradictory*, seen in the 'productivity dilemma' (Clark *et al.* 1987) of discord between innovation and productivity, change, and experience—what economists sometimes refer to as the tensions between 'dynamic' and 'allocative' efficiencies, and organization theorists, the tensions between 'exploration' and 'exploitation'. It can be *conservative* (Morgan 1986), *unreliable* (March *et al.* 1991), and *non-uniform* (Marengo 1992). Learning *from failure* is important, as is *unlearning* (Hedberg 1981). The sources of learning *change* over time (Mody 1990). The lack of preparedness of firms to fund it (Pucik 1988*b*), and problems in establishing incentives to encourage it (Aoki 1988), provide some of the many reasons why organizations encounter such difficulty in learning (Argyris and Schon 1978).

Nevertheless, firms do learn, survive, and improve comparative performance. This learning progresses beyond the 'everyday' adaptation and improvement organizations can achieve in their existing competencies through learning by 'doing' (Arrow 1962) and 'using' (Rosenberg 1982), and encompasses 'higher-level' learning (Fiol and Lyles 1985), which questions the validity of current competencies and facilitates the construction of new ones. This is described in the management literature as 'generative' as opposed to 'adaptive' learning (Senge 1990), and by Argyris and Schon (1978) as 'double-loop' and 'deutero' learning as compared with 'single-loop' learning. It is a key element of a firm's innovative capabilities.

In conditions of rapid and disruptive market and technological change, firms can be argued to need 'higher-level' learning, because existing 'lower-level' learning focuses on current systems, products, and technologies and not on new competencies and opportunities. However, such learning is constrained. Much of the management, innovation, and business economics literature points to the conservatism of the strategies firms adopt, reflecting what companies are currently best at, rather than what changing markets require. The emphasis of much of this literature is that 'history matters', and that what a firm can do in the future is strongly influenced by its past and its collective learning. Firms' learning is described as 'path dependent' (Dosi 1988), and their technological choices are constrained by their 'technological trajectories' (Dosi 1982) and 'firm-specific accumulated competencies' (Pavitt 1991). Yet, as shown in the longevity of large industrial firms, resiliently operating under very different technological conditions (Pavitt 1993), firms do learn to achieve 'higher-level' learning—that is, they learn to learn, in ways that question their assumptions about what they do, and transform their competencies.

Firms learn through purposefully adopting strategies and structures that encourage learning (Senge 1990; Dodgson 1991; Malerba 1992). They learn through the activities of key individuals or 'boundary-spanners' (Michael 1973), through executive succession (Tushman *et al.* 1986), and through recruitment and training programmes. Important in this process is the way they develop shared cultures that facilitate learning (Schein 1985). Furthermore, the management strategy literature has long pointed to the ways firms not only react to external change, but proactively seek to shape the environment in which they operate and learn (Chandler 1966; Ansoff 1968).

A major mechanism by which firms learn about technology is through their internal R & D efforts (which, as we saw in Chapter 3, have the ability to improve firms' adaptive capacities). Learning is thus, to a significant extent, a function of the size and focus of R & D budgets, and the strategies for their direction and management. Firms do, of course, learn from a wide range of other functions, particularly from marketing and production and from the iterative interactions between domains. The learning that takes place in production, in both development and implementation, is particularly important and leads Leonard-Barton (1992) and others to talk of the 'factory as a laboratory'. Furthermore, learning has both an 'internal' and an 'external' component. Given the complexity and multifaceted nature of learning, it requires a multiplicity of stimuli, processes, and outcomes. External learning is valuable in this respect. Malerba (1992) develops a typology of forms of learning: learning from doing, using, searching, interacting, inter-industry spillovers, and advances in science and technology. The latter three in this list he describes as primarily external activities and these will be examined in Chapter 7. Small and medium-sized firms can have some advantages in their learning activities. These will be discussed in the following section.

TECHNOLOGY STRATEGIES IN SMALL AND MEDIUM-SIZED ENTERPRISES

Around 99 per cent of all enterprises in Europe and the United States employ less than 500 persons, and these enterprises account for 50 per cent of employment in the United States and over 70 per cent in Europe. Small and medium-sized enterprises (SMEs) produce over 50 per cent of total manufacturing value-added in Japan. They are significant creators and partners in, and users of, technological innovations. According to some research, small firms are more efficient in their use of R & D expenditure than large firms, as measured by patent output (Wyatt 1984).

SMEs can play an important role in the development of new technologies and industries. According to Schumpeter's Mark 1 model of innovation, shown in Fig. 6.15, it is entrepreneurs taking advantage of new science and technology that generates new industries (his Mark 2 model emphasized the innovation advantages of large firms (Freeman and Soete 1997)). There are many different types of SME, with various relationships with technological innovation (Fig. 6.16). New technology-based firms (NTBFs) are those SMEs whose business is based on new technologies in information and communications technology (including software, multimedia, and Internet firms), biotechnology, and new materials. Niche strategy, technology-based firms are those firms which use technology as the basis for their competitiveness. SMEs in traditional sectors, such as furniture and retailing, can be extensive users of technological innovations, but are rarely its source.

In NTBF and the niche-strategy firms, technological mastery is the basis of competitiveness. In this regard, SMEs are disadvantaged when compared to larger firms in their access to resources and to some innovative capabilities (such as the amount they can afford to spend on search activities). However, they have strengths in some innovative capabilities, particularly those related to their behavioural advantages (Rothwell and Dodgson 1994). So, for example, it may be easier for them to undertake aligning, implementing, integrating, and

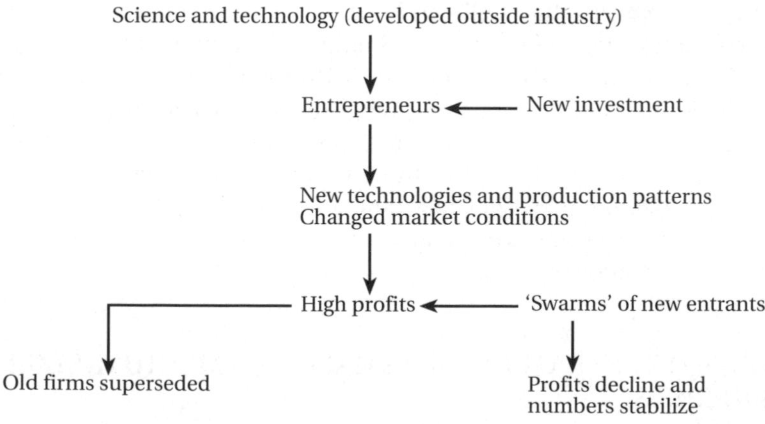

Fig. 6.15. Schumpeter's Mark 1 model of innovation

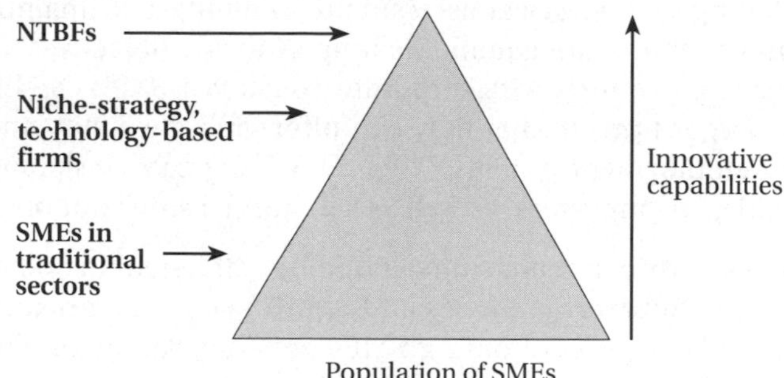

Fig. 6.16. Innovation in SMFs

coordinating functions as a result of their smaller size. They also have the capacity for fast learning (Dodgson 1991). Amongst the key advantages of SMEs is the ability to access, generate, and develop technologies more quickly and more effectively than competitors.

There are five broad factors that influence technology strategy in SMEs.

- *Accumulated technological competencies.* Firms with wide-ranging R & D expertise are better able to deal with the threats and opportunities that emerge from rapidly changing technologies. While many SMEs are started on the basis of technologies developed elsewhere, their strategic growth depends on the development of internal technological strengths. Without these internal capabilities they cannot advance the technologies they work with and successfully access novel technologies from external sources.
- *External orientation.* Small firms rarely have the technological base, or financial and staff resources, to develop and commercialize technologies themselves. They have to be highly receptive to working with other firms, research organizations, and universities, and be very effective in the management of such partnerships, in order to be competitive in the long term.
- *Organizational specialisms.* Smaller firms adopt organic styles of management more easily than larger firms. Staff are generally better able to communicate across functional and departmental boundaries (which may not exist at all in very small firms) than staff of large firms. The flexibility and adaptability of SMEs are a key competence, as is their ability for fast learning.
- *Internal strategic cohesion.* Although small firms often have difficulties in developing and implementing strategies, in successful high-technology, high-growth companies, the scale of the firm's activities, and the limited number of people determining strategic direction, enable integration of corporate decision-making across business and technology areas.

- *Management skills.* These include the range of skills found in larger firms, including technological assessment, building and maintaining benefits from collaboration, communicating strategic objectives, and integrating technology strategy with corporate objectives. SMEs can have advantages over larger firms in that they can offer great flexibility and incentives in their employment systems. They can offer greater task range and responsibilities to employees, as well as substantial stock options.

There is a body of research into technology strategies in SMEs (Dodgson and Rothwell 1991; Roberts 1991; Jones and Smith 1997). This research points to the particular challenges confronting SMEs growing rapidly on the basis of their technological competencies. For innovative SMEs to grow successfully they need to overcome some 'thresholds' in their activities and behaviours that can constrain their development.

The first threshold confronting an SME is its *start-up phase.* It is at this stage that it is determined whether the firm has the requisite resources and expertise to develop a business. Other thresholds emerge after the company has established itself and is entering into growth phases. A second threshold is *market and product expansion.* Many SMEs need to make the transformation from single to multiple products, and from customized to batch production. They often need to move from a focus on small-market niches to higher-volume, more standardized markets. As domestic markets are often too small to allow for sufficient growth, they have to export.

A third threshold facing SMEs is *technological diversification.* If an SME's technology is derived from a scientific base—for example, if it emerged as a spin-off from a university or research organization—then there is a need to master the production and engineering skills required to produce commercially. The firm needs to integrate its R & D and production activities. If, on the other hand, the technology derived from an engineering function—for example, if it emerged as a spin-off from an engineering firm that was not interested in developing a particular new product—then there is a need to understand its basic underlying principles so as to assist the development of new generations of product. It needs to undertake formal R & D.

We can gain more insight into the technological diversification threshold by considering Roberts's (1991) four levels of technology evolution, shown in Box 6.6. He argues that 'the best opportunities for rapid growth of a young firm come from building an internal critical mass of engineering talent in a focused technological area, yielding a distinctive core technology that might evolve over time, to provide a foundation for the company's product development' (Roberts 1991: 283). Long-term growth, therefore, depends on major enhancements of core technologies.

Another threshold facing SMEs concerns *management transformation.* Growing firms need to make the transformation from being entrepreneurially managed to being professionally managed. An entrepreneur is rarely the most appropriate person to run a larger company. The sorts of skills required of an

Box 6.6. ***Roberts's four levels of technology evolution***

Minor Improvements

Include new products that embody minor technological improvements to correct known problems.

Major Enhancements

Involve substantial improvements and advance of a technology in which the company has previous expertise.

New, Related Technology

Involves developing an entirely new core technology that is integrated with an existing company technology in a final product.

New, Unrelated Technology

Includes new core technology that is not combined with existing product technology.

Source: Roberts (1991)

entrepreneur are not readily applied to establishing organizational structures, delegations, financial reporting systems, and personnel policies required when a company expands.

A brief note on financing growth in technology-based SMEs

Large firms usually have a greater range of financing options for their R & D, including retained earnings, share issues, bonds, and loans. SMEs rarely have access to this range of funding, and the absence of these resources is often a major factor constricting their growth. What sorts of finance enable SMEs to overcome the thresholds described above? Roberts's (1991) studies in the United States show that the sources of start-up finance include, progressively, personal savings, family and friends, private investors, wealthy family funds, and seed venture capital funds (Fig. 6.17). Venture capital funds, and funds from a variety of financial institutions, assist the initial growth period of the firm.

Critical in overcoming the growth thresholds is the role of private investors ('business angels') and seed funds in the United States. The 'quality' of these funds is as important as their 'quantity' in assisting firms to address the thresholds. Business angels not only bring funds into the firm, but bring their expertise and experiences of managing technology-based growth. Experienced businesspeople can give valuable advice to growing firms about the need to confront the thresholds, and how to do so. They can be instrumental in encouraging the start-up to change its funding basis from debt to equity, and thereby increase the amount of available capital to assist with the transitions required. More generally, business angels

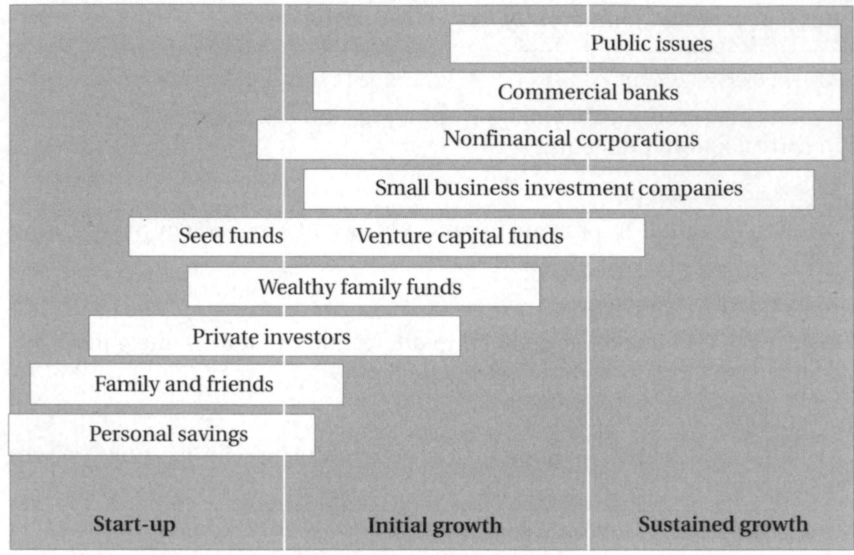

Fig. 6.17. Primary investment preferences of capital sources in the United States
Source: Roberts (1991).

- *have a business background;*
- *are willing to make equity investments that are too small for a venture-capital company;*
- *are often willing to make their experience available to the businesses in which they invest through a seat on the board of directors, a consulting role, or even part-time employment in the business;*
- *expect a significant capital gain for their investment;*
- *are willing to assess investment proposals much more quickly than equity funds and institutional investors (Chadwick 1996).*

*One estimate suggests that, in the United States, 250,000 business angels invest between $10 billion and $20 billion in 30,000 firms annually (QED, Sep. 1997). The importance of business angels can also be seen in a major UK study of small firms (*Enterprise Britain, *1998). Private individuals—typically business angels with experience and expertise to offer as well as money—financed more enterprises than venture capitalists. It would appear, therefore, that business angels in the United Kingdom also fill an important gap in funding sources for newer, higher-technology firms.*

CONCLUSIONS

Technology strategy is notoriously difficult both to practise and to analyse. Few companies have consistently managed to develop technology in a strategic manner (Christensen 1997). Without underemphasizing the complications involved in the development of technology strategy, this chapter has attempted to present a relatively simple analysis. It has presented some basic tools that companies can use to assess and manage their technology portfolios. And it has developed a simple conceptualization of what technology strategy is, and the forms it can take.

We have made a primary distinction between a firm's resources and its innovative capabilities. These are constructed in a variety of ways and, in combination, create a firm's technological competencies. The aim of technology strategy is to use, develop, and extend these competencies. This involves utilizing and investing in resources, and developing a range of innovative capabilities. These innovative capabilities include learning in a variety of ways, using different means, and at different levels.

Firms' strategic options are largely constrained by their existing resources. If a firm's competitive advantage can be maintained by the use of its existing technological competencies, then a consolidation approach can be used to ensure these resources are used to best effect. If firms' future competitive advantages cannot depend on existing resources, they can attempt to develop the innovative capabilities to change those resources. Changes in innovative capabilities are required if a firm wishes to differentiate itself through non-price factors, and when diversification is pursued. The type of innovative capability required varies with the level of diversification required.

The technological-competencies view of technology strategy emphasizes the lengthy, complex, and multifaceted nature of technological accumulation. Building resources and innovative capabilities requires substantial continuing investment. Decentralization and the reduced time horizons of R & D discussed in Chapter 3 are, therefore, inappropriate behaviours for firms wishing to build technological competencies and operate strategically with their technology.

Small and medium-sized firms face particular issues in the strategic management of technology. These often relate to issues of managing growth and navigating the various thresholds that can constrain their development. As with large firms, the success of SME strategy depends to a large extent on their capacity to learn and to this extent they have comparative advantages in their capacity for fast learning.

CHAPTER

7

Technological Collaboration

WHAT IS TECHNOLOGICAL COLLABORATION?

Collaboration, in the form of technology-based joint ventures, strategic alliances, and multi-partner R & D projects, is an increasingly important feature in the generation and diffusion of technology and is therefore a key MTI activity. Technological collaboration essentially involves shared commitment of resources and risk by a number of partners to agreed complementary aims. Vertical collaboration occurs throughout the chain of production for particular products, from the provision of raw materials, through the production and assembly of parts, components, and systems, to their distribution and servicing. Horizontal collaborations occur between partners at the same level in the production process. As we have seen in previous chapters, vertical, user–supplier links play a central role in the innovation process. Horizontal links also assist the innovation process, although firms may be comparatively more reticent to form such collaborations, as these may lead to disputes over ownership of their outcomes, such as intellectual property rights, or to direct competition between collaborating firms. Technological collaboration—with suppliers, users, partners, and universities—was a feature of all the case-study companies in Chapter 1.

Technological collaboration between firms can take a variety of forms. It may be a joint venture, formed by two or more partners as a separate company with shared equity investments. It could be a partnership linking firms on the basis of continuing commitment to shared business or technological objectives without equity sharing, often known as 'strategic alliances'. It may take the form of R & D contracts or technology exchange agreements, whereby firms' shared objectives involve the interchange of research findings or technological know-how. Increasingly, universities and public research laboratories are partners in such R & D contracts. Where such relationships abound amongst groups of firms and public-sector institutions, they are sometimes described as 'innovation networks' (Freeman 1991).

WHY DO FIRMS COLLABORATE IN THEIR TECHNOLOGICAL ACTIVITIES?

There is a wide range of explanations for why firms collaborate. They include those discussed in Chapter 2, where we examined changing industrial structures and systems of production, and more particularly the role technology and innovation play in this change. There are also economic explanations (cost reduction) and the competitive relationships between firms (competitor exclusion or locking-in key players), while other explanations are less instrumental and focus on qualitative issues like organizational learning (see Dodgson 1993*a* for a discussion of these various approaches). Generally, however, from the perspective of the firm, technological collaboration is seen as a means of improving technological competencies and learning about new markets, management practices, and strategies.

Although collaboration occurs in many different forms, and may reflect different motives, a number of generalizable assumptions underpin them. First is the belief that collaboration can lead to *positive sum gains* in internal activities—that is, partners together can obtain mutual benefits that they could not achieve independently. Such benefits may include the following.

- *Increased scale and scope of activities.* The outcomes of collaboration may be applicable to all partners' markets, and thus may expand an individual firm's customer bases (increased scale). Synergies between firms' different technological competencies may produce better, more widely applicable products (increased scope). Increasing the scale of resources to technology development can also raise entry barriers to other firms.
- *Shared costs and risk.* Collaboration can share the often very high costs, and therefore risk, of innovation (although it also, of course, shares future income streams).
- *Improved ability to deal with complexity.* Closer strategic and technological integration between firms is a means for dealing with the complexity of multiple sources and forms of technology. It allows, for example, the better transfer of tacit knowledge (see Chapter 2).

A second assumption regarding collaboration concerns the way it assists with *environmental uncertainty.* Increasingly sophisticated and demanding customers, growing competition in and globalization of markets, and rapidly changing and disruptive technologies place pressures on firms to exist with, and attempt to control, these uncertainties. This is believed to be achieved more easily in collaboration than in isolation. A number of analyses of collaboration link it with uncertainties in the generation and early diffusion of new technologies (Freeman 1991). The product life-cycle model of Abernathy and Utterback (1978), for example, implies a cyclical role for collaboration based on uncertainty. Thus, in early stages of development there are periods of high interaction between organizations, with many new entrant companies possessing technological advantages over incumbent firms, and extensive collaboration between firms until a 'dominant design' emerges in a technology. As the technology matures, uncertainty declines and collaborative activity recedes. The high level of collaborative activity seen in the creation of technical standards is a means of reducing uncertainties by introducing interchangeable products and interfaces (discussed in Chapter 8).

A third set of assumptions underlying collaboration concerns its *flexibility* and *efficiencies* compared to the alternatives. For example, collaboration may be an alternative to direct foreign investment, mergers, and acquisitions, which are much less easily amended once entered into. We saw in Chapter 6 that, as a governance structure, collaboration has advantages over the alternatives of arm's-length transactions and vertical integration. It can allow firms to keep a watching brief on external technological developments without having to invest heavily. Large-firm–small-firm interaction can be facilitated such that the resource advantages of the former are linked with the behavioural or creative advantages of the latter whilst each maintains its independence. A large drug company, for example, may choose to collaborate with our biotechnology firm in Chapter 1, as a means of developing its options, so that it could invest more heavily once the technology is better proven and better understood. The large firm will have gained the opportunity to learn about the technology during the collaboration.

While information and communications technologies have facilitated increased and more effective collaboration, much technological knowledge is not only tacit, but firm specific (Pavitt 1986). It is, therefore, difficult to transfer easily or quickly. Collaboration potentially provides a mechanism whereby close linkages among different organizations enable the development of sympathetic systems, procedures, and vocabulary that may encourage the effective transfer of technology. It may also allow partners to 'unbundle' discrete technological assets for transfer (Mowery 1988). Finally, collaboration may address the difficulty of valuing technological knowledge by providing a means of exchange that does not necessarily rely on price.

Potentially, therefore, there may be numerous advantages to be achieved through collaboration if these assumptions hold. That is not to deny the potentially adverse aspects of collaboration. Technological collaboration can be anti-competitive, by excluding certain firms, or raising entry barriers, or operating in the form of cartels that anti-trust legislation prevented in the past. Also, there may be strategic dangers from firms that overly rely on externally sourced rather than internally generated technology. Without internal technological competencies there can be no 'receptors' for external technology, nor capacity for building the technological competencies that provide the basis for firms' technology strategies (and that provide the basis for attracting potential partners).

None the less, the efficacy of collaboration as an aid to innovation can perhaps best be judged by consideration of its current extensive use. However, examining whether the potentials of collaboration are in practice being realized is difficult, as data on its extent and outcomes are often piecemeal and frequently contradictory. Furthermore, whereas the bulk of evidence suggests an increasing role for collaboration in industry, the majority of studies of its outcomes point to the considerable difficulties in gaining mutually satisfactory outcomes amongst the partners (Dodgson 1993*a*).

THE EXTENT OF TECHNOLOGICAL COLLABORATION

Measuring the extent to which technological collaboration occurs is notoriously difficult. There are, however, numerous examples from around the world of increased collaborative activity. The encouragement of technological collaboration is a key policy focus of the European Commission, as seen in policies such as ESPRIT (a collaborative programme with an IT focus) and various Framework Programmes (funded collaborative research in a range of industries). In the United States, SEMATECH provides an example of government-sponsored technological collaboration, and, in Canada, the IRAP scheme encourages collaboration between firms and universities. A wide range of technological collaborations also occur in Japan, ranging from large-scale, high-technology schemes such as the Fifth Generation Computer Project to local support schemes in over 150 Regional Technology Centres. Taiwan's Industrial Technology Research Institute has played a central role in encouraging technological development and diffusion through collaborative projects. Large Korean firms are increasingly forming technological alliances with US, Japanese, and European firms.

A number of databases on technological collaboration measure the numbers of new international alliances announced in the technical press (see Fig. 7.1). These tend to cover high-profile, technology-creating projects and under-represent more diffusion-oriented collaborations and those based outside non English-speaking countries. The best of these databases, the MERIT-CATI database, which only includes inter-firm collaborations, shows the increase in the number of new collaborations being formed throughout the 1980s and 1990s. The majority of these new collaborations occur in new technologies, particularly in IT, and are based in the United States, Japan, and Europe.

The extent to which firms source technology externally, both vertically and horizontally, is affected by industrial structures. A commonly cited reason for the high levels of external integration in Japanese industry, for example, is the structure of industry itself. The large business groups—the *Keiretsu*—which control a wide range of diversified interests and can facilitate close trading relationships and cooperation, and the strong vertical relationships found in Japan down the supply chain, with technological linkages between contractors and subcontractors, are argued to be a major source of technological dynamism (Economic Planning Agency 1990). Relationships between contractors and

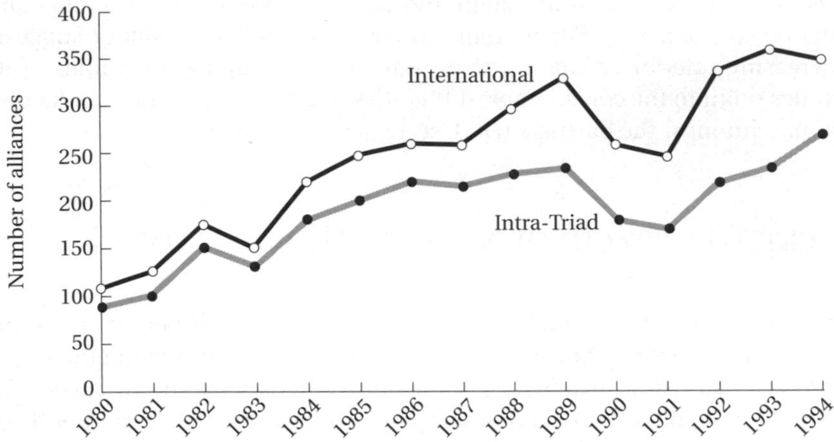

Fig. 7.1. Number of newly established technology alliances, 1980–1994
Source: Hagedoorn (1997).

their usually smaller subcontractors in Japan have changed considerably from the period during the 1950s when smaller firms were seen to be low-cost suppliers whose services were easily utilized during expansionary times and readily dispensable during recession. Since the 1990s small Japanese firms have been seen increasingly as a source of technology for larger firms (Dodgson *et al.* 1995).

In their survey of 114 US automotive suppliers, Kamath and Liker (1990) found that firms dependent upon the automotive industry (those having more than 60 per cent of their sales in autos) were much more likely than non-dependent firms to invest in innovation, even if the results were not cost effective in the short term. The importance of this relationship is also seen in Hull and Azumi's (1991) survey of invention rates in Japanese firms. They found that inter-organizational relationships with suppliers are positively associated with invention rates. They argue that major firms with dependent suppliers appear to have high invention rates, in part because improvements made by their vendors accrue to them.

THE CHALLENGES OF MANAGING TECHNOLOGICAL COLLABORATION

Firms vary in their capacity to use externally sourced technology. Mansfield (1988*a*) pointed to international differences by showing Japan's advantage over the United States in the way its firms innovate based on external technology. He found that, for technologies based on sources of knowledge outside the firm, Japanese firms exhibited significantly shorter development and commercialization cycles. In the United States firms took almost as long, and spent almost as much money, to carry out an innovation based on external technology as one based on internal technology. In Japan, on the other hand, firms took about 25 per cent less time, and spent about 50 per cent less, to carry out an innovation based on external technology than one based on internal technology. These international differences may or may not remain at the start of the new century, but they point to the possibility of substantial differences in the capacity of firms to collaborate effectively.

It is difficult to consider what constitutes success in managing collaboration, since the range of firms' circumstances, their expectations, and their experiences of collaboration are so variable. Some firms are content with satisfactory technological outcomes, others require bottom-line financial improvements as a result of their collaborative activity. In all cases, managing successful technological collaboration is a difficult task. This is so because of changes over time during the course of collaborations and because of what can be called the *technology collaboration paradox.*

The changes that can occur during the course of a technological collaboration are similar to those that can affect all joint ventures and include

- changing aims of collaboration;
- changing bargaining power of partners;
- obsolesence of the original reasons for forming the collaboration;
- initial focus on the wrong sets of issues (Harrigan, 1986; Contractor and Lorange, 1988).

The technological collaboration paradox derives from the way in which one of the greatest attractions of collaboration is the possibility of learning from partners. Organizations learn more from dissimilar organizations, where technologies, cultures, management practices, and strategies are different and where the opportunities for learning are therefore greater. A small firm may be attracted to working with a large firm because of its high levels of resources and well-established operational procedures. Alternatively, the large firm may be attracted to working with the small firm because of its flexibility and entrepreneurialism. Private-sector firms may be attracted to working with universities or research laboratories because of their relatively unfettered, curiosity-driven research, while these labs may work with large firms because of their resources and commercial expertise. In all cases it is an attraction based on varied competencies, which are themselves the result of very divergent organizations with different ways of working. Herein lies the paradox: the more attractive the partner in this sense, the greater the opportunities for miscommunication and misunderstanding because of the differences between the partners.

Partner selection

As a result of these potential problems, partner selection is the most critical decision affecting the success of collaboration. There are advantages in selecting partners for long-term relationships. As many of the motives for technological collaboration reflect attempts to deal with complexity and uncertainty in novel and rapidly changing technologies, it is perhaps unsurprising that there are advantages in partnerships with long time horizons. In long-term relationships the problems in the technological collaboration paradox may be overcome. There is greater opportunity for firms to exchange information equitably. Managers and engineers in different companies can develop better working relationships, and technologies are more easily and comprehensively transferred. If these advantages are to be obtained, selection of the partner should be made on the basis of the long-term attractiveness of the collaboration, as well as the intrinsic interest of the proposed project.

Partner selection should, therefore, be a strategic decision. As we saw in the previous chapter, one of the purposes of articulating a technology strategy is to communicate long-term objectives to potential partners. Thus, in many successful collaborations negotiations over partner selection are undertaken by top management, but this is not always the case. Some collaborations are undertaken for operational reasons and are negotiated at a decentralized level (Chen 1996). Such collaborations, however, will not produce the same level of strategic learning as that which can occur from greater corporate involvement. Success also seems more likely when partners provide complementary technologies. Complementarity in expertise is frequently cited as a reason for the technological success of the collaborations that enabled partners to learn novel skills (Dodgson 1993*a*).

Collaborations facilitate the transfer of knowledge. Once this knowledge has been transferred, the need for the partnership may be assumed to be finished. However, as technologies and markets are continually developing, the transferred knowledge may no longer be the most appropriate for changed market conditions. Individual firms continue in their efforts to develop their specialist skills through R & D. For this reason there are advantages in firms collaborating not only on the basis of existing technology, but on the understanding that partners may continue to improve their technological competencies. These improvements are related to the comparative advantages of individual firms.

Technologies may be completely complementary, but firms may have totally incompatible business aims. Collaborators will generally not want to compete in the same markets using the product of their partnership. Potential markets need to be demarcated, either on a product basis or geographically. Technology and business strategies need to be sympathetic and mutually supporting.

Scientists and engineers may find it difficult to work with people in similar positions in other firms with lower levels of competency. Specialist vocabulary may not be common, understanding of latest research techniques or findings may not be shared. Unequal competencies result in delays and diversion of efforts as the weaker partner is brought up to speed. Unless there is an element of respect for the partner's abilities, transfer is unlikely to be wholehearted. International collaboration accentuates the need for greater respect for partner competencies. Furthermore, awareness of the commensurate abilities of partners may provide a stimulus to creativity. It may, for example, provide an element of competitiveness between research teams that may assist innovation.

Flexible and adaptable structures

The process of collaboration is often described as tension ridden (Roehl and Truitt 1987). In part these tensions derive from the way technologies and markets constantly change. Unless collaborations are dynamic, they may be aiming at a target that has moved. Throughout the course of a collaboration, opportunities may arise that were initially unforeseen, and outcomes from collaboration often may not be the ones originally envisaged. For these reasons, collaborations need to be adaptable in structure and purpose.

A case study in collaboration tension

When Du Pont created a new technology venture with BT, the synergies appeared obvious. Du Pont wanted to diversify out of chemicals into electronics, and BT had some excellent electronics technology in its research laboratory that it wanted to commercialize. However, within a short period the following tensions arose.

- *Changing strategies of the partners.* Increased prices for chemicals helped Du Pont refocus priorities back into its traditional industry. BT shifted its strategy away from being a technology supplier towards becoming a service company.
- *Inexperience of one partner.* Du Pont had lengthy experience of joint venturing and working with small spin-offs. BT had very little and its expectations were too high as a result.
- *Reporting arrangements.* The spin-off company reported to a technical function in BT and a marketing function in Du Pont, and this caused some confusion.
- *Top management structure.* It took some time to appoint an independent managing director for the joint venture, and until that time the representative of one partner company was viewed with suspicion by the other partner.
- *Cultural mismatch.* One partner was essentially a research laboratory, without much market awareness; the other was a marketing organization, without much appreciation of the demands of R & D.
- *Harmonization of human resources.* Staff on secondment to the joint venture from the partners had the salaries and conditions of the parent company. These differed and caused some tensions when it was discovered how people were being differentially rewarded for doing similar jobs.
- *Adjustment of the target market.* The original business aim of the joint

venture was not realized. Du Pont pragmatically accepted this and allowed the joint venture (which was eventually successful) to develop a new business. BT had problems with this.

- *Changing expectations.* The joint venture became a very different company from that initially envisaged. Du Pont accepted this, as it was commercially attractive. BT found the joint venture did not achieve what it wanted—that is, the commercialization of some of its technology. Originally the joint venture was owned 50–50. After a number of years Du Pont bought a majority ownership (Dodgson 1993*a*).

The case study of Du Pont–BT supports the view that collaboration should build longer-term competencies as well as focus on particular products given that the latter may fail in the market or have limited life cycles. As the focus of collaboration may change over time—for example, as the project progresses nearer to the market—the skills mix of managers, scientists, and engineers needs to adapt accordingly.

Communications and human-resources factors

Good communications within and between firms are critical to the success of collaboration. Building effective communications paths into partnerships is often problematic, particularly for small firms linking into multinationals. Having established the appropriate reporting linkages, the next problem is using them effectively. Reporting unnecessary or poor-quality information may reduce the credibility of the whole system (and the collaboration). Without giving away all the knowledge and skill that made one partner attractive to the other, it is important to transfer information that is necessary to make the collaboration work. Sometimes one partner may feel that it is contributing more than the other. In such circumstances, a *quid pro quo* is needed. This may be achieved at later stages in the project, or in future projects. It may be achieved formally, through, for example, proportional allocation of intellectual property rights or equity. Or it may be done informally through the trading or exchange of information at the discretion of project managers (using the so-called favour bank).

Operating in periods when the flow of information is primarily one way requires a high level of trust in partners that the flow will in future be reversed. High trust relationships within collaborations are often based on the assumption of continuity and reciprocity between partners. In-depth studies of collaboration place great emphasis on the personal factors that enable trust to develop and collaboration to succeed. Communications depend on individuals, and are enhanced by the ability of individuals to be trusted. Managers, scientists, and engineers are trusted by their equivalents in other firms to deliver what is expected of them on time. Counterparts are trusted to be honest, and not to impart false or misleading information. Trust is particularly important when there is imbalance in contributions to the collaboration. The partner may be trusted to rectify the imbalance in the future. As interpersonal trust between individuals in different organizations is likely to be affected by labour mobility or individual disagreements, successful collaboration often depends upon the extension of interpersonal trust to inter-organizational trust. This issue, and broader considerations of trust, will be examined in greater detail in a later section.

The management of human resources provides the critical factor determining the success of partners within collaborations. Human resource management is an important aspect of collaboration in a number of respects. First, collaboration requires very good project managers. Such personnel need to be attracted into the partnerships, without, for example, jeopardizing career and pay prospects by working in what might be seen as a subsidiary activity. Secondly, given the importance of interpersonal communications in collaboration, the retention of key individuals—managers, scientists, and engineers—is crucial. Thirdly, attention to human-resource issues can reduce the tensions that sometimes occur in collaborations because of the lack of harmonization between the salaries and conditions of partners.

ORGANIZATIONAL LEARNING AND TECHNOLOGICAL COLLABORATION

There is a literature that places learning centrally in its analysis of inter-firm links (see, for example, Contractor and Lorange 1988; Doz and Schuen 1988). For Kogut (1988*b*) joint ventures are 'vehicles by which knowledge is transferred and by which firms learn from one another'. These approaches suggest that learning provides motive for and desired outcomes from inter-firm links. Thus, learning is necessary to comprehend and respond to changing industrial and technological systems. It can assist in developing competencies and enhance the power that provides competitive advantage. Inter-firm links also assist firms' internal constraints to learning. An external orientation assists firms to overcome the organizational introspection described in the management and organization theory literature, and applicable to firms' R & D groups. Psychologists refer to the way people learn *vicariously*, by watching before they perform and profiting from the successes and failure of others (Bandura 1977). Firms similarly can use external links to learn in this way. External links can bring new knowledge into the firm of a specific, project-based nature. They can also enable firms to reconsider their existing ways of doing things—be it in R & D organization or the implementation of new technology. If successful, external links may lead to a realignment of business strategy through diversification (Ciborra 1991). Inter-firm links provide an opportunity to observe novelty through the approaches of partners, can stimulate reconsideration of current practices, and can be an antidote to the 'not-invented-here syndrome'.

Learning vicariously can also help prevent the repetition of mistakes, and inter-firm links can provide opportunities for 'higher-level' learning (Dodgson 1993*c*).

The primary ascribed *motive* for learning through inter-firm links is to deal with technological and market uncertainty (Mody 1990; Ciborra 1991). Ciborra (1991: 59) argues that alliances are the institutional arrangement that most efficiently allows firms to implement strategies for organizational learning and innovation. 'The alliance brings into the corporation new expertise concerning products, marketing strategies, organizational know-how, and new tacit and explicit knowledge. New management systems, operating procedures and modifications of products are the typical outcomes of this incremental learning.' Ciborra's analysis argues that the *outcomes* of alliances are reduced uncertainties by means of improving predictability of technological development and that they are a means of reducing the transition costs of firms' transferring strategies. These transition costs can be argued to reflect the learning rigidities in firms. Firms do become defensive, introspective, and resistant to change. Inter-firm links provide opportunities for expanding learning horizons and for overcoming internal introspection (Dodgson 1993*c*).

Consideration of organizational learning also reveals the *process* of inter-firm links. Such consideration is important for a number of reasons. First, because it is necessary to account for adaptability and change in inter-firm links. As Contractor and Lorange (1988) argue, the strategic rationales prevailing when a cooperative venture was formed may shift over time. They argue that the erosion of the fundamental strategic rationales may come from external or environmental sources (such as technological obsolescence) or internal sources, 'such as when one partner learns from the other, and the other partner has nothing to contribute'.

Secondly, the differential speeds at which partners learn have marked consequences for the outcomes of inter-firm links. In his study of US–Japanese joint ventures, Pucik (1988*a*: 80) argues that 'benefits are appropriated asymmetrically due to differences in the organizational learning capacity of the partners. The shifts in relative power in a competitive partnership are related to the speed at which the partners can learn from each other.'

Thirdly, consideration of the process of linkage implies examination of the qualitative nature of relationships within and between firms. It is the process by which externally derived knowledge, both tacit and codified, is diffused throughout the organization (for example, from central to divisional R & D, from R & D to production, from individuals to groups), and hence provides benefits and returns to that learning. An insight into the need for diversity in learning sources is provided by Di Maggio and Powell (1983), who refer to 'organizational fields' of organizations, which in aggregate constitute a recognized area of institutional life (suppliers, consumers, regulators, and so on that produce similar services or products). They argue that, 'once disparate organizations in the same line of business are structured into an actual field . . . (by competition, the state, or the professions), powerful forces emerge that lead them to become more similar to one another' (Di Maggio and Powell 1983: 148).

This tendency towards 'isomorphism' has profound implications for learning. If firms in a network extensively share knowledge over a long period, then they will increasingly come to resemble one another, with detrimental consequences for novelty and innovation. Just as Marengo (1992) argues the need for diversity of learning within firms, it would appear beneficial if firms were to seek heterogeneous sources of external learning, with perhaps long-term, intimate links to assist incremental improvements, and some short-term links with companies to assist radical, higher-level learning.

It is possible to discern a variety in the approaches taken towards learning. In some, technological links are purely a means of accessing valuable technical information. In others, it is possible to see technological links as a means to enhance the ability of firms to learn. When Matsushita collaborated with a small Scottish firm, OWL, engineers were sent to work at the smaller firm in order to immerse themselves in a different, and potentially more creative, small-firm work environment. The ways these engineers' horizons were broadened were potentially valuable stimuli to learning in Matsushita (Dodgson *et al.* 1995).

Even when problems emerge between collaborators, learning occurs. For example, a small English firm, Peritas, had difficulties operating in a fast-moving environment with a larger company, VAG, and its engineers became very frustrated at continually having to change specifications. However, at an end-of-project review to find out what had been learnt throughout the process of the project, it was agreed that it had been a useful learning experience. The company had learned about the technology and the process of collaborating with large firms. Similarly, small software company Attica had problems in its collaborations with large firms, but acknowledged that its link with large computer company ICL had taught the company about management systems. Despite some unsatisfactory experiences of collaboration, taking a broad strategic management perspective of learning from such collaborations reveals the benefits of partnerships with other firms (Dodgson *et al.* 1995).

Learning is exchanged not only between the partners in a particular collaborative project, but also between the broader networks of participating firms. One of the key requirements of successful collaboration is for participants to understand the nature, process, and likely outcomes of partnerships. While learning from experience is likely to be the major form of instilling such knowledge, it is noteworthy that the approach taken by the Japanese government's Plaza Program eases firms into collaboration by a lengthy 'getting-to-know-you' procedure, before joint projects are established. The cautious approach to forming partnerships and the recognition of the need for a high level of trust between collaborators is a feature of the Japanese public-policy approach to forming technological linkages. This is in stark contrast to the approach sometimes seen in European firms of rather opportunistically forming links between often disparate partners in order to receive R & D grants.

TRUST AND TECHNOLOGICAL COLLABORATION

The quality of relationships between partner firms has obvious implications for the outcomes of technological linkages. A wide number of studies show how effective inter-firm links and learning between partners depend on high levels of trust (Jarillo 1988). Lundvall (1988: 52), for example, argues that, in order to overcome the inevitable uncertainties in jointly developed product innovations, 'mutual trust and mutually respected codes of behaviour will normally be necessary'. Saxenian (1991: 430) contends that 'a network of long-term, trust-based alliances with innovative suppliers represents a source of advantage for a systems producer which is very difficult for a competitor to replicate. Such a network provides both flexibility and a framework for joint learning and technological exchange.' The firms in this network are argued to exchange sensitive information concerning business plans, sales forecasts, and costs, and have a mutual commitment to long-term relationships. This involves 'relationships with suppliers as involving personal and moral commitments which transcend the expectations of simple business relationships' (Saxenian 1991: 428).

Hakansson and Johanson (1988: 373) describe a range of these commitments and bonds

> Interaction between firms develops over time. It takes time to learn about each other's ways of doing and viewing things and how to interpret each other's acts. Relations are built gradually in a social exchange process through which the parties may come to trust in each other . . . Over time, as a consequence of interaction, bonds of various kinds are formed by the parties. There may be technical bonds which are related to the technologies employed by the firms, knowledge bonds related to the parties' knowledge about their business, social bonds in the form of personal confidence, administrative bonds related to the administrative routines and procedures of the firms, and legal bonds in the form of contracts between the firms. These bonds create lasting relationships between the firms.

The importance of cultural affinities within such combinations of inter-firm linkages is strongly emphasized within the 'industrial districts' literature (Hirst and Zeitlin 1989). Freeman (1991: 503) argues that 'personal relationships of trust and confidence (and sometimes of fear and obligation) are important both at the formal and informal level . . . For this reason cultural factors such as language, educational background, regional loyalties, shared ideologies and experiences and even common leisure interests continue to play an important role in networking.'

A number of reasons can be suggested for why high trust facilitates effective inter-firm links, both horizontal and vertical. The first relates to the sort of *knowledge* being transferred. It may be tacit, uncodified, firm specific, and commercially sensitive. It is, therefore, not readily transferable, requiring dense, reliable, and continuing communication paths. Furthermore, it is often *proprietorial.* What is being exchanged is the kind of knowledge and competence that is not easily replicated or purchased by competitors and thus can provide important elements of a firm's defining competencies and competitiveness. Not only are partners expected to share trust in each other's ability to provide valid and helpful responses to uncertainty; they are expected not to use this information in ways that might prove disadvantageous to their partners.

A second reason relates to the time scale of successful inter-firm links. Trust facilitates continuing relationships between firms (Arrow 1975). Continuity is valuable because, as we have seen, the objective of inter-firm links may change over time—in line, for example, with changing or new market and technological opportunities. Furthermore, it is only within a long-term horizon that reciprocity in collaboration can occur. At any one time, one partner will be a net gainer in a collaboration. The disincentive to cut and run is based on the view of future gains, which can only be achieved through continuity of collaboration. Trust mitigates against opportunistic behaviour (Buckley and Casson 1988)—as does fear of mistrust on the part of future new partners should a firm behave in such a manner.

Many user–supplier links have surprising longevity (Hakansson and Johanson 1988). This enables effective communication paths to develop, and facilitates the social and other links to be established. Macaulay's classic (1963) article arguing that contract law is often ignored in business transactions is revealing in this regard. He argues that trading partners' primary motives are to remain in business and that they will avoid doing anything that might interfere with this. This includes avoiding the legal system, but also being sensitive to reactions of business partners and concerned about their business reputation. A priority for business, he argues, is the need for flexibility over the long term. He points to the way detailed negotiated contracts can get in the way of good exchange relationships. 'Some businessmen object that in . . . carefully worked out relationship(s) one gets performance only to the letter of the contract. Such planning indicates a lack of trust and blunts the demands of friendship, turning a cooperative venture into an antagonistic horse trade' (Macaulay 1963: 64).

The advantages of cooperativeness and fairness in continuing reciprocal relationships are demonstrated by Axelrod (1984). In his analysis of the large variety of approaches to the 'prisoner's-dilemma' game,[1] he castigates the unsuccessful contestants ('expert strategists') from political science, sociology, economics, psychology, and mathematics for their 'systematic errors of being too competitive for their own good, not being forgiving enough, and being too pessimistic about the responsiveness of the other side' (Axelrod 1984: 40).

[1] The prisoner's dilemma is as follows. Two people are held on suspicion of a crime and prevented from communicating with each other. They know that, if they both confess, each will get four years' gaol; if neither confesses, both will still be convicted of part of the crime and will get two years. However, if one confesses and the other does not, the former will go free and the latter will be gaoled for five years. The study of this 'non-zero-sum' scenario has focused on mathematical analysis of the relationship between 'strategies' (e.g. always confess or never confess) and outcomes, over many repetitions of the game.

Axelrod argues that his approach, based on the assumption of continuing interactions, has implications for the conduct of business relationships.

A third reason for the advantages of high trust in collaboration reflects the high management cost of such linkages. Selecting a suitable partner and building the dense communications paths through which tacit knowledge can be transferred has considerable management costs, both real and opportunity. These costs are increased when consideration of interpersonal trust is extended to inter-organizational trust. Trust between partner firms is commonly analysed by means of relationships between individuals. Given the problems of labour turnover and the possibilities of communications breakdowns on the part of particular managers, scientists, and engineers, to survive, trust relationships between firms have to be general as well as specific to individuals. They have to be ingrained in organizational routines, norms, and values. Inter-organizational trust is characterized by community of interest, organizational cultures receptive to external inputs, and widespread and continually supplemented knowledge among employees of the status and purpose of the links (Dodgson 1993*b*). Such features are not costless, and, once the effort has been made to build such strong relationships, jeopardizing them through a lack of trust is not a sensible option.

INTERNATIONAL TECHNOLOGICAL COLLABORATION: JOINT VENTURES IN CHINA

China has been involved in joint ventures with Western business partners since the 1970s, and their incidence increased following the 1984 urban reforms. Many of the early joint ventures—manufacturing jet engines, aircraft, and other equipment—were largely military focused, and, while there were some successes, in common with experiences of international joint ventures in general, there were a number of big failures. Despite the problems with joint ventures, they are a common feature of international business, and there is now a wide range of research available on the policy and management issues related to international joint ventures (Mytelka 1991), and some systematic research on joint ventures in China (Newman 1992; Williamson and Hu 1994).

The Chinese government has a preference for joint ventures as a means of acquiring technology from overseas. Mansell and When (1998) identify four reasons for this preference.

- The Chinese partner has a say in the management of the joint venture.
- The commitment of the foreign investor to make the project succeed is secured.
- The transfer of sophisticated technology requires close interaction.
- The transfer of much-needed managerial know-how is included.

A range of factors affect the formation, conduct and process, and outcome of joint ventures in China.

General Motors in China

General Motors (GM) planned in the 1990s to increase its production capacity in the Asia Pacific by 25 per cent, with manufacturing and assembly plants in China, Thailand, Indonesia, and India (Chan 1997). According to the executive secretary of the GM–China Technology Advisory Board, Tai Chan, GM's new business relationships in China 'require cooperation on technology—which will enable us to meet local requirements—not only regulatory or business-related requirements, but also societal and consumer-related needs and preferences'. Along with GM's Delphi Automotive Systems and Delco Electronics business units, GM has seventeen continuing business relationships in China.

GM has a partnership with the Shanghai Automotive Industry Corporation to build mid-size sedans in China. The partnership was approved by the Planning Council of China after a long selection process, and was based in part on GM's advanced automotive engineering capabilities. As part of the agreement, technology institutes have been set up in Shanghai and Beijing in conjunction with the vehicle programmes, and GM operates a Pan Asia Technology Automotive Center, which acts as a technology integrator for vehicle design and development in the joint venture. GM's R & D Operations coordinates the GM–China research portfolio, with research being undertaken in six Chinese universities, and a number of state laboratories and technology joint ventures. The GM–China Technology Advisory Board also transfers automotive technology and scientific and engineering knowledge through formal and informal exchanges such as conferences (Chan 1997).

The formation of joint ventures in China

There is a variety of motives for encouraging joint-venture creation in China (Box 7.1). There are considerable differences in the motives of different participants underlying the promotion of, and strategies for, joint ventures, which can, of course, lead to major difficulties. Whatever the motives for choosing to form joint ventures, the most critical decision affecting outcomes is partner selection. The selection process encompasses

- the number, and size, of partners;
- the location of the proposed venture;
- choosing partners with complementary assets and strategies to ensure immediate and long-term synergies;
- finding partners with sympathetic investment aims regarding ownership of equity and expected financial returns;
- a clear understanding of the regulatory constraints likely to be faced by the joint venture, including those pertaining to foreign currency exchange.

Selecting a partner is a lengthy and difficult process. It may take several years, and requires decisions on the most appropriate way of proceeding with

Box 7.1. ***Motives for joint-venture creation in China***

Chinese government motives	International company motives	Chinese company motives
Improving industrial efficiency	Accessing new markets	Accessing new technology
Reducing government intervention	Accessing new materials	Accessing new skills; in production, organization and quality control, and general management
Promoting the development of the private sector	Accessing labour-cost advantages	Cash injections
Reforming state-owned enterprises without recourse to full privatization	Accessing investment capital	Avoiding domestic controls in wages, regulations, and political interference
Developing capital markets	Difficulties with other forms of foreign investment	Accessing special regulations and tax arrangements
Reducing fiscal deficits	Learning from local partners	Prestige of international operations
Creating jobs	Internationalization strategies	Accessing foreign markets
Broadening public ownership		
Attracting foreign investment		

selection—whether to go it alone or use intermediaries (and, if the latter, how these are selected). Locational decisions are affected by the availability of incentives in particular regions, cities, or special economic zones. Williamson and Hu (1994) describe the difficulties in choosing a location, suggesting that decisions need to be based on

- avoiding 'walled cities' (i.e. entrenched competition);
- pre-empting input supplies and sites;
- levering off traditional skill bases;

- benefiting from customer proximity;
- accessing ancillary industries and infrastructure.

They emphasize the difficulties in forming joint ventures by arguing that these factors almost inevitably pull the selection decision in opposing directions.

Conduct and process

A number of US studies have revealed the very different approaches to management of joint ventures adopted by Americans and Chinese. Some of these differences are highlighted in Baird *et al.* (1990), whose attitudinal survey reveals how Chinese managers show a significant preference for an impersonal, less participative, and more individual style of joint-venture management. Americans, by contrast, favour a personal, more participative, more team-oriented and consensual approach. Essentially, the greatest difference is the Chinese preference towards the impersonal, rather than the personal approach.

This finding is confirmed by Newman (1992), who suggests that the modern style of highly flexible, participative, and democratic management in the West cannot be transferred without modification into a Chinese setting. Based on US experience, he recommends a 'focused joint-venture' model that first defines and sticks closely to prescribed ways of performing activities. The second characteristic of his model is to select a narrow product line. The third essential element for success is the maintenance of strong support by both partners. Fourth, is the motivation of Chinese employees to overcome social and cultural barriers to becoming personally committed to the venture. Fifth is the maintenance of consistent product quality and delivery from local Chinese sources of supply. Finally, Newman raises the issues of long-term planning or the need to change strategic direction in the course of an agreement as areas of potential difficulty between Chinese and Western partners.

Additionally, it is important to consider the negotiation of rights and obligations, not only with the partners but with local and central governments. The issues here include taxation, employment (including training requirements and redundancy arrangements), occupational health and safety, supply of services (water, electricity), and adequate maintenance of infrastructure. Particular problems are likely to emerge over the transfer of technology: what is to be transferred, how it is to be achieved, and what is to occur when the transfer is completed—that is, when one partner has learned all it needs from the other. Governance structures need to be created such that each partner feels it has sufficient and appropriate control. Alternatively, systems need to be put in place for management training and the replacement of expatriate with local managers such that equity in governance can be achieved in the future.

Outcomes

Measuring the outcomes of joint ventures is compounded by the problem of defining success and failure. Partnerships often work in some areas, but not others. One partner may be satisfied with outcomes, another deeply unhappy. Expectations can differ widely. For example, in China a successful outcome may be defined by employment or export growth rather than profitability. Studies of the factors that lead to success and failure in joint ventures are greatly needed, particularly to address the problems of differences in motivation for forming them, differences in approaches to their conduct, and differences in assessment of their outcomes. The following case study of a ten-year technological collaboration between a British and Chinese company shows how many of these problems were overcome.

Dalian Locomotive Works–Ricardo Engineering joint venture

Dalian Locomotive Works (DLW) in China and Ricardo Engineering in the United Kingdom began working together in 1980 to help improve DLW's locomotive engines. DLW is one of China's three major locomotive manufacturers. It employs 12,000 workers, and it has excellent engineering skills. Ricardo is one of the world's premier engine design companies. It employs 1,700, and has seventy-five years' experience in engine R & D. Three project stages were undertaken over the first fifteen years of the DLW–Ricardo relationship.

First, a DLW engine was assessed by Ricardo. This involved

- design and performance evaluation;
- recommended design improvements aimed at a 20 per cent improvement in fuel consumption and durability by introducing a turbocharger and improved fuel injection;
- a full functional specification for an improved engine.

Secondly, a detailed analysis and test was undertaken, involving

- designs produced by complementary CAD/CAM systems and concurrent engineering;
- an engine being shipped to Ricardo;
- Ricardo focusing on fuel injection and turbocharging, DLW concentrating on single cylinder testing;
- DLW manufacturing components;
- rig testing at both partners;
- DLW producing a prototype;
- successful testing and move to full production.

Thirdly, there was a comprehensive redesign of major components, involving redesigns to increase power and performance.

The technological outcome of the collaboration was a substantially improved engine, achieved more quickly and at lower cost than DLW could have achieved by itself. The partners also learned a great deal technologically, organizationally, and strategically. Ricardo learned about Chinese engineering practices and standards; DLW mastered the technology of its engines and developed its own design capabilities. Both partners improved their international and cross-cultural management skills. Ricardo developed the strategic skills to give it the confidence to do business in China, and DLW increased its technological self-sufficiency.

There were tensions throughout the relationship. Particularly important were difficulties over negotiations, language, and engineering culture.

- *Negotiations.* Ricardo found it very difficult to negotiate with both the firm and the Chinese bureaucracy in the Ministry of Railways, each of which had different motives for working with Ricardo. There were also cultural differences in approaches to negotiation, with the Chinese being very thorough (the three projects, for example, involved twelve months of negotiations). Negotiations were always difficult, given DLW's very limited foreign currency, and its need, from Ricardo's perspective, to invest heavily in CAD/CAM equipment.
- *Language.* Forty DLW employees learned English during the collaboration; some Ricardo employees learned a few words of Chinese. There were instances of miscommunication that led to problems, such as when the Ricardo team reported successful outcomes, having met 99 per cent of target aims. DLW felt this constituted failure, as all aims were not completely met.
- *Engineering culture.* Ricardo engineers were schooled in a tradition where individual judgement was encouraged. DLW relied more heavily on comprehensive tests and data. There was occasional mistrust in the early stages of the relationship, when Ricardo wanted to proceed based on individual opinion, and DLW insisted on comprehensive and definitive data.

DLW and Ricardo worked hard at resolving these tensions, and did so through

- establishing mutual trust and respect at senior manager and individual engineer level;
- consistently agreeing to common objectives, sticking to contracts, being prescriptive about performance and highly specific about planned outcomes;
- Ricardo learning at various levels about dealing with Chinese Ministry bureaucracy and about the social importance of meetings and etiquette;
- DLW learning about the culture of the individual engineer;
- regularly rotating DLW staff through Ricardo to assist technology transfer;
- carefully managing personnel changes in Ricardo, so that there was continuity and maintenance of trust;
- openly discussing problems and delays.

The case of DLW–Ricardo shows that successful outcomes in joint ventures often depend upon effective procedures for either conflict resolution or conflict avoidance through careful management.

CONCLUSIONS AND SUMMARY

Firms collaborate in their technological activities for a variety of reasons, including a belief in collaboration's capacity to deliver positive-sum gains, deal with environmental uncertainty, and deliver flexibility and efficiencies compared with alternative forms of organization. The extent to which technological collaboration occurs is hard to gauge, but some databases show it to be growing in importance, particularly amongst the triad countries. This is occurring despite firms experiencing a great deal of difficulty in managing collaboration well. Collaboration can have other negative connotations, including its potentially anti-competitive aspects. Furthermore, over-reliance on collaboration and its use as a substitute for in-house R & D is a flawed strategy. Such an approach reduces the capacity of a firm to receive knowledge from outside sources and to adapt externally derived technology to its particular needs, and restricts the development of technological competencies.

The greatest management challenge of collaboration is the selection of partners. There are many advantages in choosing partners for the long term, and ensuring that the technologies each partner contributes are complementary. Because the process of technological collaboration is so difficult and prone to unexpected developments (both positive and negative), it is advantageous to use flexible and adaptable structures for management. Key elements of successful collaborations include high levels of interpersonal and inter-organizational trust, and efforts to maximize the organizational learning that can occur.

International technological collaborations can be fraught with difficulties. The potential benefits and problems of international collaborations can be seen by examining the development of joint ventures in China. These joint ventures, which often have a technological focus, face profound differences amongst the partners in their motives, and views about their conduct, process, and outcomes. A case study of a lengthy joint venture between a British and a Chinese joint venture illustrates the positive benefits that can be achieved if very careful consideration is given to its management.

CHAPTER

8

The Commercialization Process

WHAT IS THE COMMERCIALIZATION PROCESS?

The ultimate aim of MTI is to improve the competitiveness of firms. This involves the variety of ways that all investments in MTI—in R & D, new product development (NPD), operations and production, technology strategy, and technological collaborations—are commercialized by successfully producing income streams. When firms are successful at commercializing technology, as we saw in Chapter 1, not only does this lead to competitive advantages, but the technology can become the defining aspect of the firm and its future development. The key aspects of the commercialization process discussed here include marketing technology products, intellectual property rights and know-how, licensing, technical standards, and some issues of technology transfer.

Investments in MTI can be commercialized not only through the marketing of tangible products or services, but by the protection of intellectual property (which provides firms with opportunities and options for the future) and their subsequent sale through the medium of licensing, and through the establishment of technical standards. There are a number of problems with licensing, particularly concerning the pricing of technology, and these will be discussed. Ownership of technology also produces value for a company in other ways, such as increasing stock prices and company valuations (Pfizer's Viagra is a case in point). The commercialization process sometimes also requires the transfer of technology—transferring it to where it can be used profitably—and this will be examined briefly.

MARKETING TECHNOLOGY PRODUCTS

The primary means by which technological investments are commercialized is through the production of products, components, and services, which are then sold in the marketplace. We have seen, however, that a high proportion of new products fail in the market. It is the marketing domain that has the task of reducing these high levels of failure, and some of the ways it does so have been discussed in Chapter 4. Marketing expertise would have made a significant difference to the biotechnology company in Chapter 1. Whilst its technology was excellent, the company was unaware of how best to use it, and it originally targeted the wrong market.

Previous chapters have emphasized the importance of direct inputs from marketing into the broad range of firms' technological activities. The high levels of internal and external organizational integration required by MTI include close links between marketing, R & D, and operations and production, and their further integration through technology strategy. Marketing plays an important role in the use of stage-gate systems, where it disciplines the process of NPD towards considerations of market needs. We also saw how, in the case of the Post-It notes, marketing can make some serious miscalculations, occasionally preventing firms from realizing opportunities. Poor integration of marketing and production input into the NPD process can lead to the sort of difficulties Du Pont experienced when it developed Kevlar. The original market for the product was to be tyre cord, a large market, but the company had not properly understood the costs of production, and the costs of alternatives. It took some time before the product was successfully used in other applications.

There are three major roles for marketing in the commercialization of technological investments through technology products and services. First, market definition, or posing the question: what should we make? The marketing function has a critical role to play in defining what R & D to undertake, what new products to develop, and what sort of production technology is required. Littler (1994: 295) argues that marketing's functional

> role is seen to be concerned with commissioning and/or undertaking market research and analysis, and with having an active part in the development of all aspects of the offering that include pricing, advertising, promotion, service support, distribution, packaging, sales and design. Its prime purpose should be to ensure that the offering which emerges from the development process has significant appeal to the customer segments which it has identified as having the optimum potential for the business, whilst at the same time having a perceived differentiation from its competitors with regard to those values which its customers regard as important.

Customers, both industrial and individual consumers, can often be segmented into groups with specific requirements, and marketing can assist in articulating, defining, and measuring these requirements.

Secondly, and relatedly, marketing plays an important role in facilitating internal and external communications. According to Littler (1994: 294)

> the marketing function may also have a key part in gathering, analyzing and disseminating throughout the organization intelligence on customer purchasing behaviour, satisfaction levels, attitudes towards the business and its competitors and such like, as well as contributing to the development of an overall corporate culture which not only acknowledges the central role of the existing and potential customers but also the manner in which the dynamics of the environment are continuously shaping demands,

resulting in new customer priorities, with consequent implications for the development of new and existing products.

Littler argues that firms market not just a product, but rather a collection of values such as the ability to perform tasks, enhance appearance, and augment or reinforce perceived self-image. When it comes to technology products, where there is an increasing commodification of technical product features, it is the non-technical features, such as service quality, distribution, and technical support, that add the greatest value. He argues that there is a temptation for technology producing firms to concentrate too much on the features of the technology, and it is the role of marketing to ensure that new products satisfy the basic criterion of presenting the customers with something that they regard as having some differentiating benefits, such as ease of use.

Thirdly, there is a field of marketing that concentrates on issues of relationships between firms and their management (Hakansson 1982). We have seen how close customer–supplier relationships are important in the development of new products, inasmuch as longer-term, more intimate relationships engender the trust required to exchange sensitive and valuable information. The loss of a major relationship with a customer in such circumstances can be very difficult, so it is important in such cases for the marketing department to be involved in the conduct of the relationship. It is also important to ensure that firms do not become too attached to particular clients, and have the capacity to diversify their sources of information about customer needs (Hakansson and Snehota 1995).

A case of successful international technology marketing: Netafim

Netafim is an Israeli firm that manufactures drip irrigation products. In 1997 it achieved 90 per cent of its $210 million sales from exports. The company was set up in 1965 by some agronomist farmers in a kibbutz, and its drip irrigation product was developed to deal with problems kibbutz members faced in irrigating their own crops.[1] For the fiftieth anniversary of Israel's independence, a team of experts chose Netafim's product as the most important Israeli invention since the state was founded.

Israel has a strong scientific base in agriculture (its milk cows, for example, are claimed to be the most productive in the world). Israeli agriculturists have pioneered agricultural biotechnology, soil solarization, and the sustained use of industrial waste water for agriculture. There is a well-established extension service system, bringing research results quickly to field trials, and problems are brought directly to the scientist for solution (Israel, Ministry of Foreign

[1] In drip irrigation a perforated plastic pipe is laid on the ground. The perforations are designed to release a controlled amount of water near the root of plants. The method minimizes water losses due to both evaporation and deep seepage below root level.

Affairs, 1999, www.mfa.gov.il). Making optimal use of scarce water, harsh land, high salinity, and extremely hot summers has led to a focus on water-saving techniques.

Netafim has extended its range of products to include a full line of drippers and dripper lines, micro-irrigation equipment, mini-sprinklers and sprinklers, landscape irrigation equipment, peripheral systems, and turnkey project services. By 1995 it had four manufacturing plants, producing five billion drippers annually. By 1995, over 20 billion drippers were in use worldwide, and Netafim drip systems had been supplied to more than eighty countries.

The company invests over 5 per cent of annual sales on R & D, with a focus on improved system efficiency and broadened applications. New products are designed by the Product Research and Development Department in Tel Aviv, while production R & D is the responsibility of R & D departments at each production centre. Product development is updated via research undertaken by the Agronomic Research and Training Departments, and by feedback on product performance from local and export sales departments. Between 1994 and 1996 the company invested over $28 million on production capacity. Its production lines are fully computerized, and operate twenty-four hours a day. The production machinery and software were exclusively designed by the company (see http://www.netafim.com).

The effort and resources the company invests in technology are matched by its commitment to marketing. Netafim's international marketing and service operations have a worldwide network of subsidiaries, local agents, and distributors. It has subsidiaries in Australia, Brazil, China, the Czech Republic, France, Germany, Italy, Korea, Morocco, Mexico, Philippines, Russia, Poland, South Africa, Thailand, and Zimbabwe. The company is also reputed to have (indirect) sales in Iran. Its marketing network also sells the products and services of Netafim's affiliated companies. It provides comprehensive technical and agronomic support services to export customers, and to agricultural and landscape experts. Company agronomists, geologists, soil and plant experts, water engineers, and other support are frequently sent out from Israel, and local agents participate in regular updates and training sessions. Each country is served by a desk of specialists, who regularly visit their customers and understand local conditions (www.netafim.com).

Netafim is building a base in India and China, both large countries with enormous water problems and demand for efficient irrigation systems. In India it is building a subsidiary factory that will bring production closer to the market, save on transportation, and also take advantage of subsidies available to local producers (which can be as high as 50–70 per cent). It has also embarked on a $40 million agreement in China to build an irrigation systems plant, and to undertake various projects. One of the projects will install advanced irrigation systems in high-tech greenhouses in a desert region of China, and will be overseen by China-based Netafim employees.

Netafim provides an excellent example of a company whose growth has resulted from a combination of substantial investments in product and process innovations with extensive technological marketing and support activities.

INDUSTRIAL (INTELLECTUAL) PROPERTY RIGHTS AND KNOW-HOW

The move to the knowledge economy, as we saw in Chapter 2, increases the importance of commercializing intangibles such as intellectual property rights (IPR). Annual patent registrations in the United States exceeded 200,000 in 1996, compared with around 120,000 in the 1980s (*The Economist*, 1 Dec. 1997). Technology licensing and royalty payments increased from $7 billion in 1976 to over $60 billion in 1995 (World Bank 1998). Intellectual property earnings from overseas in the form of royalties and licence fees grew as fast as all service exports between 1993 and 1996 in the five largest OECD economies (UK Department of Trade and Industry, 1998). Many firms put enormous value on the ownership of intellectual property, protected by patents. Table 8.1 provides a list of the major patentors in the United States in 1998.

Some companies, like Texas Instruments, which received nearly $700 million in royalties in 1995, rely heavily on income from intellectual property. The ownership of intellectual property in the twenty-first century can be likened to the ownership of physical property (land) in pre-industrial societies. It is an important basis of commerce, and ownership indicates wealth and achievement. With good management its owners flourish and those without it are highly dependent and excluded from important aspects of business. Rather than selling tangible products, firms can commercialize their technological investments by selling their intellectual property.

TABLE 8.1. *International top ten patenting companies in the US patent system, 1998*

Company	Patents
IBM	2,682
Canon	1,963
NEC	1,725
Samsung	1,553
Motorola	1,426
Sony	1,351
Hitachi	1,322
Toshiba	1,271
Fujitsu	1,260
Matsushita Electrical Industrial	1,239

Source: CHI's Research, Apr. 1999.

Failure to protect intellectual property can have major commercial implications. Cultivation of the popular Kiwi fruit was developed following extensive R & D in New Zealand. Failure to protect this research or to register the trademark now sees the successful production of this fruit in South Africa and Latin America, with the resulting loss of potential markets for New Zealand. We have seen the problems faced by IBM when it failed to control important technologies.

The characteristics of intellectual property and know-how, however, are very different from those of products, as shown in Box 8.1. Because of these characteristics, if commercialization is to occur, and firms are to enjoy returns to their investments, intellectual property has to be protected. Industrial property laws recognize and confer certain exclusive rights on the proprietors of industrial property rights. There are various categories of IPRs: patents, petty patents, industrial designs, trademarks, know-how, and trade secrets (the definitions that follow are based on Blakeney 1989).

- *Patents* are 'a statutory privilege granted by a government to an inventor and to other persons deriving their rights from the inventor, for a fixed period of years, to exclude other persons from manufacturing, using or selling a patented product or from utilizing a patented method or process'. Patents are usually awarded for between fifteen and twenty years, and in some countries can be extended. After this time the invention is in the public domain. Patent laws vary internationally, but usually patent applications must
    - contain a description of the invention, and all drawings referred to,

Box 8.1. ***A comparison of intellectual and physical property***

	Know-how	Physical entity
Publicness	Use by one party does not prevent use by another	Use by one party prevents simultaneous use by another
Physical depreciation	Does not wear out	Wears out
Property rights	Non-existent for many elements	Generally exists for most items
Protection from encroachment	Title-holder often cannot exclude encroachment	Title-holder can generally exclude encroachment
Transfer costs	Increased with the tacit portion	Depends on transporta-tion costs

Source: Teece (1998).

disclosing it in a manner sufficiently clear and complete for it to be carried out by a person skilled in the art;
  - be presented in the context of the state of the art;
  - provide a novel solution to a technical problem, involve an inventive step (a 'creative advance on existing knowledge'), and be industrially applicable (non-theoretical).

The registering of patents can be extremely complex, and often requires the professional skills of a patent attorney. The process of registration involves the Patent Office examining the patent to see if formal requirements have been met, then examining its substance. This involves searches of other patents and may involve public inspection before the patent is granted. When the patent is granted, it is published and laid open for public inspection. The patentor can add subsequent patent improvements or additions.

- *Petty patents* are awarded for more limited technological advances than those that are patentable. They have a shorter duration.
- *Industrial designs* are defined as 'the ornamental or aesthetic aspect of a useful article'. These are registered as a design patent or certificate of registration. They must be novel and repeatable in commercial quantities.
- *Trademarks* are 'a sign which serves to distinguish the product of one enterprise from the products of other enterprises'. A trademark must be visible and can include names, existing or invented words, letters, numbers, pictures, symbols, and even sounds.
- *Know-how and trade secrets.* Know-how is the way in which technology is applied in a commercial situation. It can be protected only if a restriction is placed on its unauthorized communication as a provision of a technology transfer agreement or if the know-how is considered to be confidential. Most common-law systems imply a contractual term obliging employees not to disclose information that is considered to be the property of an employer.

Strong intellectual property law not only provides incentives for firms to innovate by providing temporary monopoly positions, it also encourages firms to export their technologies. A World Bank study found that the strength or weakness of a country's system of intellectual-property protection has a substantial effect, particularly in high-technology industries, on the kinds of technology that many US, German, and Japanese firms transfer to that country (World Bank 1998).

A brief note on international law on intellectual property rights: Trade-Related Aspects of Intellectual Property Rights (TRIPs)

Intellectual property rights are created by national law and thus apply only in a single national jurisdiction, independent of such rights granted elsewhere.

Establishing a global IPR regime thus requires cooperation among national governments to harmonize their separate laws. Numerous international treaties to promote such cooperation have been negotiated over the past 100 years. Most are administered by the World Intellectual Property Organization (WIPO), a specialized agency of the United Nations. WIPO conventions—for example, the Paris Convention for industrial inventions and the Berne Convention for copyright of literature, art, and music—require their signatories to grant national treatment (foreign firms are treated in the same way as domestic ones) in the protection of IPRs, but typically do not impose common standards of protection. New global rules on IPRs are forcing a reassessment of past strategies for acquiring, disseminating, and using knowledge.

The 1994 TRIPs agreement builds on existing WIPO conventions and lays the foundation for global convergence toward higher standards of protection for IPRs. It requires signatories to apply the principles of national treatment and most-favoured nation (MFN) status to intellectual-property protection. Unlike most other international agreements on IPRs, the TRIPs agreement sets minimum standards of protection for all forms of intellectual property: copyright, trademarks, service marks, geographical indications, industrial design, patents, layout designs for integrated circuits, and trade secrets.

In each area the agreement defines the main elements of protection: the subject matter to be protected, the rights to be conferred, and the permissible exceptions to those rights. For the first time ever in an international agreement on intellectual property, the TRIPs agreement addresses the enforcement of IPRs by establishing basic measures to ensure that legal remedies are available when infringement occurs. Disputes between WTO members over TRIPs obligations are subject to the same dispute settlement procedures that apply to other WTO agreements.

The provisions of the TRIPs agreement became applicable to all signatories at the beginning of 1996. Developing countries were granted a four-year transition period, except for obligations pertaining to national and MFN treatment, an additional five-year transition for product patents in fields of technology not protected before 1996 (this applies to pharmaceutical products). The least-developed countries were granted a transition period extending until 2006, again excepting national and MFN treatment (World Bank 1998).

An intellectual property company: the case of the Orbital Engineering Company

The Orbital Engineering Company (OEC) is a business based on intellectual property. Originally its strategy was to become a manufacturer of engines, but this vision was not realized. Instead it managed to create a viable business through the careful management of patents and licensing.

The company was founded by Ralph Sarich in West Australia in 1969 to develop an orbital engine. The highly innovative orbital engine possessed an orbital rotor guided by four specially shaped cranks in a multi-compartment combustion chamber. The engine was believed to be significantly more efficient and lightweight than existing automotive engines. However, some complex technical problems arose concerning seals and it became clear that auto engine manufacturers were not prepared to retool their manufacturing plants to produce the engine (the engine was not 'designed for manufacture' (see Chapter 5)). The orbital engine was never developed commercially.

Using some of the related technology developed during the design of the orbital engine, OEC turned its attention to a more conventional two-stroke engine. It produced a new engine, the OCP, which OEC hoped would revolutionize the US car industry. This obviously did not occur. While many car manufacturers, such as GM, Ford, Toyota, Honda, Volkswagen, Jaguar, Renault, Peugeot, Volvo, and BMW assessed OEC's technology, they were not convinced of the advantages of two-stroke engines. OEC's engine did not allow the car manufacturers to deal fully with increasingly stringent emissions regulations, and improvements in existing four-stroke engine technology limited the potential advantages of the new technology.

The history of OEC, and the role of Ralph Sarich and his family, is somewhat controversial. At its peak, around 1992, the company employed nearly 400 people and made a pre-tax profit of over $20 million. Subsequently, however, staff levels were reduced to around 300 and annual financial losses were common during the mid-1990s. The company received substantial government support, and investments from a major Australian company, on the basis of a number of speculative engineering opportunities, many of which were never realized. The founder and his family, having undertaken significant risk, gradually sold all of their shares, making a substantial profit. The situation in 1999 was strongly positive. The company's technology was being integrated into a significant proportion of the outboard-motor market, and was being used in the development of scooters and in a small number of experimental cars. The company's financial circumstances appear promising (Sykes 1998).

How did the company make the transition from being product-based to intellectual-property based? It is interesting to speculate on whether the transition would have occurred so readily had the original entrepreneur remained with the company (although, according to some, he did display considerable pragmatism when it came to such changes (Sykes 1998)). Using the terms of the analyses presented in Chapter 6, this question could be reframed as whether the company would have made the transition from an entrepreneurially to a professionally managed company. The basis of the transition was the use of some of its technology related to the orbital and OCP engines, particularly its expertise in direct fuel injection (df/i) and some proprietary testing equipment. Using these technologies (with the df/i proving particularly attractive to engine manufacturers), it developed a strong patent position. It registered over 100 patents in twenty-one countries, spending up to three-quarters of a million dollars annually on patent protection. Its patenting strategy involved registering an inner and outer ring of patents. Registering such a 'family' of patents has been found to award greater protection, and is commonly practised by large Japanese companies. The outer patents were used to provide an early warning of potential competitor infringements. The integrity of its patent position is illustrated by the case of Ford, which examined the patents for two years before acknowledging that it needed to negotiate a licence (Manley 1994). OEC's patent position is further strengthened by the fact that its major investor was Australia's largest company, and could bring its resources to bear in any potential litigation.

Having protected its intellectual property, the company used a sophisticated approach to attract licensees (Willoughby and Wong 1993; Manley 1994). This approach included the following characteristics: an 'approach plan' for each target firm, gauging the interest of key individuals before making the corporate contact, establishing the nature and location of organizational power centres, anticipating idiosyncratic questions, and researching the position of the potential client in respect of OEC technology. Its licensing strategy involved minimizing the decision-making time of potential clients and maximizing the impression of competitive interest, securing total technology licences as opposed to licences for individual components, high fee payments (up-front fees for major automotive companies wishing to study the technology ranged from $20 to $30 million), non-exclusionary licences, agreement not to depart from essential OEC designs in later R & D, and an agreement to provide R & D feedback (Willoughby and Wong 1993; Manley 1994). OEC's licensing and contract engineering income totalled around $100 million in the six years between 1990 and 1995, and the company was expecting substantial licensing income from Mercury Marine from 1998 (Sykes 1998).

LICENSING

As we saw in the case of OEC, valuable income streams from IPR can be derived through the use of licensing if they are carefully managed. Pilkington's float-glass manufacturing process, developed during the 1950s, brought substantial licensing income into the firm well into the 1980s. A licence is the grant by an owner of IPRs, or the proprietor of know-how, of permission to make use of all or some of those rights and information.

Firms license out their IPRs so as to exploit their existing competencies and technological assets. According to Bidault (1989), these include market advantages, production advantages, and technology advantages. Potential *market* advantages derive from the use of licences to open up new markets for a firm's technology and, in doing so,

- use local knowledge;
- avoid overseas and domestic marketing costs;

- assess market viability;
- sell semi-finished products or parts that are not 'strategic' (forcing the licensee to import key components);
- may assist diffusion of the licensor's technical standards;
- may give products a *local* image, to please consumers and governments.

Potential *production* advantages derive from the use of licences to improve the cost or quality of supplies and, in doing so,

- avoid expensive overseas manufacturing;
- avoid domestic manufacturing;
- may use the comparative advantage of licensee (in technology or natural resources);
- overcome supply constraints of governments;
- may extend product range (if the licensee has more extensive production capabilities);
- prevent the licensor from having to acquire additional capacity itself.

Potential *technology* advantages derive from the use of licences to receive income or access to other technologies and, in doing so,

- may involve peripheral technology the licensor does not want to exploit itself, but licensing can also involve core technologies;
- discover whether the technology may have applications in other markets where the licensor has no experience.

Firms purchase licences for a number of reasons. These include

- overcoming the problems associated with the absence of R & D capacity;
- avoiding much of the cost, delay, and risk of R & D;
- enabling the acquisition of complementary technology;
- improving the quality of existing products;
- launching new products;
- increasing efficiency;
- penetrating new geographical markets;
- taking advantage of market protection from government, and its information provision capacities;
- speeding up learning and accumulation of know-how;
- building on existing relationships.

Systems of cross-licensing are common in the pharmaceuticals and electronics industries. These usually occur when both partners have a similar level of technological competence; instead of the firms resorting to financial payments, technology is paid for with another of equal value.

There are numerous problems with technology licensing (Bidault 1989). These mainly derive from the differential access of the parties to information. A licensee, for example, may not know the real cost and potential of a technology on offer, and may choose overpriced or inappropriate technology. A licensor may be concerned with the prospect of the licensee becoming self-reliant once the technology has been transferred, and reneging on the licensing agreement. Another problem concerns the question of technology pricing, and this will be examined below.

A technology transfer company: the British Technology Group

The British Technology Group (BTG) was established in 1981 as a public-sector corporation.[2] It was privatized in 1992. Its major business is the development and exploitation of technology drawn from public-sector sources (universities, research council establishments, and government laboratories) and brought by private-sector firms into industrial production under licence. Essentially, BTG manages the technology transfer process, with particular reference to the protection and reward of IPR. In 1999, 75 per cent of its income derived from business outside the United Kingdom.

It has had marked successes. It assisted, for example, in the development and exploitation of magnetic resonance imaging, the Hovercraft, and Pyrethrins (which in the 1990s gave it 25 per cent of the world agricultural insecticide market). It also played an important role in the establishment of Celltech, Europe's leading dedicated biotechnology firm, and, by extension, commercial biotechnology in the United Kingdom (Dodgson 1990*b*). It has also faced criticism, however, for its failure to patent a particularly important development in biotechnology.

BTG claims to be the largest technology transfer organization in the world. Its activities are diverse—it deals with new products, processes, and computer software inventions and know-how—and highly international. Despite a post-privatization rationalization of its portfolio, by 1993 it had protected over 1,500 inventions and had 500 licensees worldwide. Of the more than 10,000 inventions protected by BTG and it predecessor the National Research Development Corporation, less than 800 have made money and of these only a dozen have provided million-dollar-plus returns. At least two, however, have made over $100 million in licensing income.

BTG employs 180 people, many of whom are scientists and engineers experienced in both academia and industry. It receives over 500 inventions a year and, historically, has proceeded with around one-third of them. The acceptance of an invention requires a favourable assessment of its commercial potential by an executive in one of BTG's operating divisions. All these executives have an academic and research background as well as managerial and commercial experience in their field of technology. Once BTG undertakes commercially to exploit an 'invention', it assumes responsibility for all patent actions and costs. BTG's patent department is one of the largest in the United

2 This section is based on Dodgson and Bessant (1996).

Kingdom. Once it has obtained a patent, BTG aims to identify industrial companies that may be interested in using the invention and negotiates licences. Finding appropriate licensees is assisted by its database of over 6,000 companies' technological interests. Licensing agreements vary in complexity but usually involve an initial down payment (to deter those not seriously interested). Royalty levels are commonly around 5 per cent for patented technology, although these may rise to 50 per cent for software.

Guaranteed annual royalties are often written into agreements, particularly where exclusivity to manufacture and sell in particular regions, usually for a limited period, is a feature. Minimum licence royalties are also negotiated to ensure that licences are effectively used and not bought just to prevent the development of potentially disruptive technologies. BTG protects inventions on a speculative basis, at no cost to the inventor or his or her institution, in anticipation of future licence income, which it shares with the inventive source on a 50/50 basis once BTG has met its patent and legal costs. The inventor receives the first $7,500 of any licence income.

BTG has a number of strengths within the context of industrial innovation.

- Preparedness to take the long-term view. It often supports projects with a 10–15-year time horizon. It has the 'critical mass' and ingrained tradition to persevere with technologies despite knowing how long they often take to turn into commercial products and processes.
- Great depth of knowledge and experience in the protection of IPRs. It has detailed knowledge of patenting systems around the world, particularly in the United States. On one occasion it fought a lengthy and successful legal battle in the United States with the Pentagon over infringement of Hovercraft patents, and it boasts of its US reputation as a 'vicious patent litigator'. The scale of its activities enable BTG to compete with universities' own patenting organizations. It is much more successful than universities in dealing with larger projects.
- Provision of technical, legal, commercial and patent resources. Until 1993 BTG also provided financial resources through investment in companies. Following privatization, however, since the major shareholders are venture capitalists, it decided instead to refer firms to its shareholders for their financial requirements.

Increasingly, BTG is developing its business of searching out technology developed by firms but not being used by them. It contends that large firms typically utilize only 20–30 per cent of their inventions to sustain their businesses. Through its Intercorporate Licensing Division, BTG licenses technology that the enterprises concerned would not themselves commercialize. It also has a growing US subsidiary and has provided a model for a technology transfer organization that has been emulated in many European countries and in Japan, Korea, and Australia.

TECHNOLOGY PRICING

There are a number of difficulties with remuneration for the sale of IPR, rather than products. First, unlike a product, IPR is intangible and decisions about its value are often more speculative and can sometimes be assessed by only a few people (mainly scientists and engineers). Secondly, the IPR one wants to purchase is at the same time the information that is needed to make a rational decision as to whether or not to buy it. This is described by Vaitsos (1974) as 'the irony of knowledge'. Furthermore, there is a danger that the information necessary to inform a potential licensee of the value of IPRs may be sufficient for that firm to gain enough knowledge to proceed independently, so the pricing regime occasionally has to account for that risk. The dangers of premature disclosure (which can destroy the value of IPR) places additional emphasis on the need for strong patent protection.

Two broad forms of remuneration are used in licensing: fixed and variable. *Fixed* payment is a negotiated lump sum, paid in one or more instalments. This form of payment guarantees the level of income to licensor, and places most risk on the licensee. *Variable* payment is made proportionately according to an agreed scale by means of a *royalty*. Royalties are very common and are paid annually or quarterly as a proportion of sales or units sold (and, rarely, profit). The licensor in this system takes most of the risk (success depends on the licensee), but potentially can receive high profits if the licensee is very successful. Much depends upon good risk assessment and belief in licensee competencies. For the licensee, this system implies reduced risk, as payments are spread over time and depend on success. However, success implies substantial payments, and there is often a commitment to continuing scrutiny by the licensor.

To overcome some of the uncertainties, firms usually use *variable-rate* royalties with a sliding scale of payments, depending on success (often determining a maximum and minimum level) or a system of lump sum plus royalties repayment—as used by BTG. As in the case of OEC, to overcome the 'irony-of-knowledge problem', when it is necessary to reveal technical and commercial information to demonstrate benefits, firms ask for *disclosure* fees. These fees are sometimes deducted from royalties. *Option fees* are sometimes paid by licensees for delaying the signing of agreement for a time while further research is undertaken, before the 'option' to buy is taken up (Bidault 1989).

When technology is difficult to price, there may be advantages in cross-licensing, or collaboration (as we saw in Chapter 7). Exchanging information over the course of a number of projects should end up in balance (in the estimation of participating scientists and engineers).

TECHNICAL STANDARDS

Another method assisting the commercialization of technology investments is through the establishment of technical standards. Firms that are involved in establishing standards have the advantage in that their design and production facilities are already geared up to meet the technical requirements. Should they decide to do so, other firms will have to meet those requirements in order to compete.

Technical standards can be established by standards authorities, such as the International Organization for Standardization (ISO), by voluntary agreement within an industry, or may exist *de facto* in line with the standards of predominant companies. The vast majority of standards, however, are not *de facto*, and involve negotiation between companies, academics, standards authorities, and other government departments.

Standards are very important for the development and the diffusion of a new technology (David 1986). In telecommunications, for example, the introduction of global standards (the GSM standard) has created a common base for designing and adapting products. It has also created a global market. Sales of products no longer depend on various domestic standards, and as a result new mobile phones and value-added services can be sold globally, thereby increasing the scale of production and reducing prices. Since the 1970s there has been a rapid growth in the number of technical standards. They are argued to have a wide range of benefits, including:

- reduction of transaction costs, by improving recognition of technical characteristics and avoidance of buyer dissatisfaction;
- provision of physical economies by simplified design, production economies, and ease of service;
- advantages to buyers through interchangeability of suppliers, better second-hand markets and spare-parts suppliers, and enhanced competition for sellers; increased product innovation (except in the case of *de facto* standards) (Reddy *et al.* 1989: 18).

Government programmes for collaborative R & D can play an important role in the development of technical standards. As Mytelka (1991*a*: 197) argues,

> firms in dynamic knowledge-intensive industries see the need to develop standards at a far earlier stage in the production process. Collaboration in precompetitive R & D of the sort promoted through ESPRIT has a particular advantage in this respect because it ensures an early approach to the harmonization of technical solutions—environments, architectures, interfaces. At a later stage in the investment process, common standards enable firms to develop compatible products and this, it is argued, is a prerequisite to the creation of new markets in these dynamic sectors.

Reddy *et al.* (1989) argue that the nature of standardization activity varies across stages in technology life cycles. Thus, activity in the emerging stages of a technology focuses on the creation of a common language, 'nomenclature and symbols'. The early stages also begin to address performance expectations, inspection, testing, and certification procedures. As 'dominant designs' in the new technology emerge, standards focus on dimensional and variety reduction. The process of standard formation is never static and continues to be evaluated and revised throughout a product's life.

When technological development is very uncertain, and there is a wide range of conflicting approaches to technical and market problems, there are many advantages for companies participating in the standards-forming process. In the view of Reddy *et al.* (1989: 18), 'product standards, by their very nature, mediate and create interdependencies in industrial markets. In their role as mediators, products standards and process of standardization introduce an element of technological and operating stability into industrial markets.'

In the computer business, the move to 'open systems' could be seen as a means of reducing the influence of proprietary standards. However, according to Morris and Ferguson (1993), the opposite is true. They argue that competitive success depends upon the establishment of proprietary architectural controls, which serve as the platform for developing product families. The architecture—the complex of standards and rules around a technology—can be developed within a network of firms, with each firm controlling particular standards. By providing stability in a rapidly changing environment, proprietary architectures in open systems are argued to be in the interest of the consumer, and will become increasingly critical as technologies, such as computers, telecommunications, and consumer electronics, converge.

Standards also have an anti-competitive element. This is obviously the case in *de facto* standards, where control is exercised by single companies, but there are also examples of collusion on the part of firms and governments to provide exclusionary standards. Lamming (1992), for example, provides the example of Prometheus, the collaborative European 'intelligent highways' research programme, which is designed to develop a technical standard that will exclude competitors (non-European firms will have to comply with the standard, and this will take time and allow European firms an advantage).

As standards have competitive value, their formation is often complicated and controversial. Competing standards may exist, as in the case of analogue and digital high-definition television. Parallel standards can exist side by side. Bessant (1991) describes the way the German car industry at one time used five different standards. The evolution of standards can be a lengthy and highly competitive process. It is, furthermore, an extremely complex process. At present the standards-making bodies in information and telecommunications technologies include over ten major standards groups, and many hundreds of working-level groups. The European telecommunications standards body, ETSI, for example, has twelve technical committees, fifty sub-technical committees, and several hundred working parties. The ISO has over 7,000 standards in information technology (Dodgson 1993*a*). That firms see fit to participate in so many committees reveals the commercial importance of technical standards.

TECHNOLOGY TRANSFER

The commercialization process involves getting technology to where it can most profitably be used, and this often involves the transfer of technology.

Technology transfer can be defined as the movement of technological capability—typically a package of artefacts, information, rights, and services—from supplier(s) to potential user(s). These transfers can occur *internally* between two organizations under the same financial control, *quasi-internally* between joint venture and alliance partners, and *externally* between independent buyers and sellers. We have already discussed the processes of transferring technology within laboratories and between the various functions of the firm (Chapters 3–5). And we have examined the transfer of technology between collaborating partners (Chapter 7). Here we shall focus on external transfers and briefly examine some methods by which firms can import technology from international sources, and some tools by which firms and research organizations can analyse and manage the transfer of their technology.

Accessing international technology

Apart from actual purchase or through the result of direct foreign investments, firms can access international technology in a number of ways.

- *Reverse engineering* is a very common method of technology transfer, and was instrumental in the development of Japanese industry. It involves disassembling goods, learning about how they work and are made, and developing *improved* versions sold under the firm's own name.
- *Pirating* where replicas sold as originals require analytical and manufacturing competencies to be developed.
- *Original equipment manufacture* (OEM) is very common in electronics and consumer goods. OEM involves a local firm producing a finished product to the specification of a foreign buyer (commonly a large Japanese or US consumer electronics firm). The foreign firm then markets the product under its own name. OEM sometimes involves the foreign firm in the selection of capital equipment and managerial and technical training, and can involve close, long-term technological relationships. It provides a valuable learning experience in design and manufacturing and has been instrumental in assisting the technological development of Korea, Taiwan, and Singapore (Hobday 1995). There are some disadvantages of OEM inasmuch as the junior partner is subordinate and dependent upon the technology, components, and market channels of the large multinational firm (which often imposes strict conditions). Furthermore, the local firm is denied access to large post-production value-added, and cannot develop brand image and international marketing expertise.
- *Turn-key plants* involve the transfer of usually complex production facilities, whereby the foreign firm takes responsibility for project management, the selection of overseas and domestic suppliers, and training of plant managers and technicians.

A brief note on the importance of personnel transfer

Personnel transfer provides the most important form of technology transfer. In the United States, high job mobility moves knowledge from firm to firm. Silicon Valley firms are exemplary here. In Japan, job-rotation transfers knowledge within firms (see Chapter 4). This also occurs when employees are seconded from firms to collaborative research institutions. In Korea and Taiwan large numbers of professionals returning from the United States bring important knowledge with them. Many Asian scientists and engineers have returned from the United States because of the 'glass ceiling' that can prevent movement from research and development into management positions, and this has considerably assisted the development of technology in these countries.

Tools for analysing and managing the process of technology transfer for commercialization

A large number of tools and techniques are used to assist firms and research organizations in analysing and managing the process of technology transfer for commercialization. Two of these will be described here, selected because they illustrate some important general principles.

Jolly (1997) develops a 'Commmercialization Map' in which he suggests there are five key stages in commercializing technologies—imagining, incubating, demonstrating, promoting, and sustaining. Each of these has a definable transfer gap—in interest, technology transfer, market, and diffusion—through which the technology must pass if it is to be commercialized (Fig. 8.1). The stages involve finding solutions to a variety of technological, production, or marketing problems, and the 'bridges' between the stages involve the mobilization of resources to deal with them.

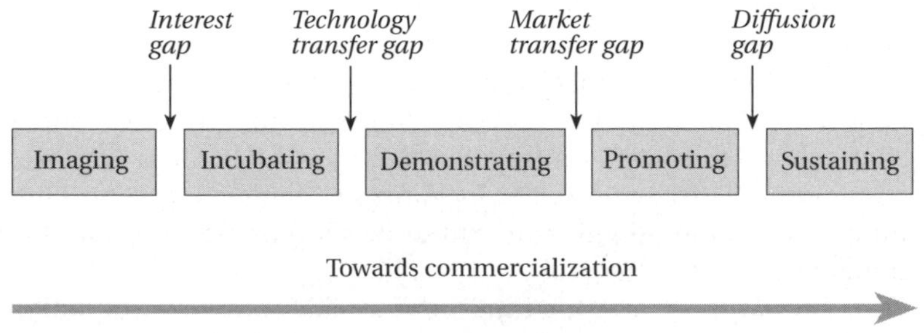

Fig. 8.1. Commercialization map
Source: Jolly (1997).

- *Imagining.* Conceiving of the technology and linking it to a market need. A vision is created (with colleagues and partners), a concept is proved, and patent protection is sought. If there is sufficient interest in the idea, in developing it, funding it, and buying it, the technology progresses through the interest gap.
- *Incubating.* The idea is fully demonstrated, technically and in a business sense, often with customer involvement. If the research originated in a public-sector institute, then this stage usually represents the end of its involvement. If the process is complete, it moves to a product development process (described in Chapter 4), and it has progressed through the technology transfer gap.
- *Demonstrating.* The first commercial quantities are produced, and suppliers and customers involved in the development are integrated into supply lines and marketing channels. Once this occurs the product has moved through the market transfer gap.
- *Promoting.* This entails careful market positioning and targeting to ensure that the product quickly gains a profitable share of the market. If this is done successfully, the product has moved through the diffusion gap.
- *Sustaining.* Here the aim is to entrench the product as broadly as possible in the market so as to ensure continuing long-term income streams.

This approach reveals several important principles.

- Commercialization can fail at any one of the stages or gaps in the process. The process throughout is highly uncertain and risky.
- The commercialization process is continuing. It does not end when the product reaches the market. The product itself may be subsequently improved, and the market may change. Long-term income streams depend not only on careful market entry strategies, but on continuing market development activities.
- As we have seen throughout this book, the commercialization of technology requires high levels of organizational integration, in which human factors, such as teamwork, are critically important.
- Integration with external organizations is also important. Early feedback from customers and opinion-formers, sound links with suppliers and, if it is not self-funded, the procurement of suitable sources of external funds, are all required.

The second tool we will discuss here focuses on the first stage of 'imagining' the new product. Called 'Quicklook', it is a methodology used at the NASA Mid-Continent Technology Transfer Center (NMCTTC) to provide a preliminary assessment of the commercial potential of a new technology. The aim of a Quicklook investigation is to provide a snapshot of the market's receptivity for a new invention (Box 8.2). Quicklook is an example of a structured, preliminary examination of the potential for a new technology. It has three basic advantages. First, it is relatively quick. It does not involve detailed market surveys, but

Box 8.2. ***Assessment of commercial potential: Quicklook***

- *Understand the technology and identify potential applications.* Typically the inventor is the best source of information about technology benefits, strengths and weaknesses, and potential applications. Inventors' insights into the aims of the research project and its history can be valuable in assessing intended and other potential applications. Staff also identify products similar to the applications envisioned during this process and, with the inventor, explore relative advantages and disadvantages of the new technology's application.
- *Identify end-users and potential licensees.* Once interesting applications and potential target user groups have been identified, it is necessary to identify appropriate and knowledgeable individuals to interview who are representative of users, developers, and manufacturers. Databases, such as the CorpTech register, Thomas Register, and Dunn & Bradstreet, make it relatively easy to identify the customers in prospective markets and manufacturers of similar products for the market. Experience suggests that R & D and marketing managers in companies that manufacture similar products are often useful interviewees. They tend to be knowledgeable and responsive, and have broad-ranging insights into the difficulties and potential of the technology.
- *Contact experts and companies.* This step is simply to call the identified contacts. Interviews with both marketing and R & D managers are valuable for overviews of the marketplace and for insights into similar products or potentially competing research projects in progress, respectively. Some types of questions interviewers might ask include the following.

  - ♦ Would a product that had these performance characteristics be important?
  - ♦ Is there a large market for products like this? Who would use it?
  - ♦ Are there similar products on the market? Who makes them?
  - ♦ What is an appropriate price point?

  Quicklook is designed as an input to the decision of whether or not to invest further in the technology rather than as an in-depth market analysis report for potential licensees.
- *Write the report.* Research findings should be compiled in a report that is easy for business managers to analyse. To facilitate comparisons between different products, managers have found it necessary to have a common format with common information about each market and technology. Each section needs to answer specific questions about the commercial viability of the product and achieve specific goals. At NMCTTC, the following topic headings are used to structure the data.

  - ○ *Technology description.* This section describes the important technical attributes of the invention in language a non-expert understands.
  - ○ *Potential benefits.* This section describes the benefits of the technology, not just the features, and the problems that the technology can solve.
  - ○ *Potential commercial markets.* This section answers the following questions.

- What products or processes could result from the technology?
- What are the health and future of the industries that constitute the market for the technology?
- What are the key technology benefits being sought by the buyers in the market?
- What are the potential market size and demand as a function of time?

- *Market interest.* This section summarizes the level of interest from those contacted.
- *Development status of the technology.* This section indicates whether the technology is a prototype, paper idea, a bench model, or other.
- *Competing technologies and competitors.* This section answers the following questions.

- What other technologies are currently being used to solve the problem addressed by the subject technology?
- Who uses similar technology?
- Does the subject technology have a demonstrable and sustainable advantage over competitive technologies in the marketplace?
- What are some of the competing companies and do they dominate the marketplace?

- *Barriers to market entry.* This section attempts to outline barriers and keys to market entry.
- *Recommendations.* The final section of the report recommends a 'go' or 'no-go' decision and outlines the steps that must be taken to help commercialize or license the technology. As a final output of the Quicklook process, Technology Opportunity Sheets are developed for external use. These are non-confidential descriptions of the technology with non-technical descriptions of its benefits and uses. They outline potential commercial uses, development status, and patent status. Based upon the Quicklook research, these have been used successfully to pique the interest of potential licensees.

Source: Cornwell (1998).

systematically collects information about the potential of the technology in a market. Secondly, it focuses the inventors of the technology, and people who use similar technologies, on its potential market. Thirdly, it enables comparisons, so that funding decisions can be made across competing projects.

CONCLUSIONS AND SUMMARY

The commercialization process is an essential element of MTI. Continuing input from the marketing function is critical to the commercialization process. Marketing informs firms about the commercial potential for technological innovations, and helps direct those innovations towards meeting commercial objectives.

Intellectual property rights provide a means by which firms can appropriate value from their technological investments. In the knowledge economy, where knowledge is extensively traded, IPR provides the security underpinning firms' transactions. Licensing is a particular form for selling IPR, but there are difficulties in managing the sale of licences for both the licensor and licensee, particularly in respect of technology pricing. The examples of BTG and OEC show that these difficulties can be managed. Technical standards can also play an important role in the commercialization process.

Technology transfer in the commercialization process can be managed through the application of various analytical tools. Two illustrations of these tools—Jolly's Commercialization Map and NMCTTC's Quicklook—have been described. Such tools help and discipline what is often a complicated and vexatious issue for private- and public-sector organizations.

CHAPTER

9

Five Future Challenges for the Management of Technological Innovation

We have seen that MTI is a complex and risky aspect of contemporary business, and how it is absolutely necessary for firms to master it if they are to compete successfully now and in the future. Various tools and techniques for the successful management of MTI have been suggested, but, as the management of MTI is an essentially idiosyncratic process, there are no simple recipes for success. The five composite case studies in Chapter 1 showed the wide variety in firms' markets, technologies, and competencies, and that the way they manage their technological innovation is directed by the different challenges they face and the overall strategies they pursue.

Some of the approaches and methods that have been discussed are relatively simple prescriptions of things to do, like using a formal-product-development management process. Others, like trying to improve organizational integration, often require extensive organizational change and are very time-consuming and complicated. Some of the tools suggested have been straightforward and practical. Others have been more analytical, particularly concerning technology strategy, which is the most difficult aspect of MTI. These practical and analytical tools will not be relevant to all firms in all circumstances; it is a question of selecting which are the most appropriate to each firm's specific situation.

One of the major arguments of this book has been that MTI, and the use of the range of analyses and methods discussed, needs to be considered in the broad context in which the firm operates. Building the most efficient factory, or developing the world's leading technology, is commercially useless if the factory is in the wrong place, or the firm is producing the wrong things, or if the technology cannot be protected, or suppliers are substandard, or it has no market. MTI has to be based on a good understanding of the broad changes occurring in underlying technologies and markets, in globalization, industrial structures, and best management practices. In an era of global competition and rapid technological change, if these changes are not understood by a firm, then this knowledge will eventually be used by competitors to its disadvantage.

The uncertainties that confront technological innovation in the early 2000s are unlikely to diminish in the future. They are, if anything, likely to increase as the pace of technological change quickens and the extent of globalization increases. Chapter 2 discussed the new contextual challenges of MTI, and subsequent chapters described how firms are changing in response to these challenges. Based on this discussion, the book will conclude by speculating about five specific management challenges for the future: technology-based competition in the knowledge economy, the new innovation process, relationships with government, science and basic research, and the globalization of science and technology.

MANAGING TECHNOLOGY-BASED COMPETITION IN THE KNOWLEDGE ECONOMY

The pace of technological change will continue to be rapid. It follows that there will continue to be an extended range of new business opportunities. As we saw in Chapter 2, the fifth wave of broad historical technological change, or techno-economic paradigm, based on information and communications technology, continues to have a profound impact on contemporary industrial organization and management. As each 'revolution' is in full swing, the next wave of technological change is in genesis. The next historical wave of technological change has been speculated to be based on biotechnology, 'green' environment, and space technologies (although we should remember the problems of predicting scientific and technological developments). Whatever the key factor industries of the future, the extent of the technological changes, and the concomitant broader changes in society and business, are unlikely to be less disruptive than in the past.

As we saw in Chapter 2, Kodama (1995) analyses the ways technologies are increasingly 'fused' across industry boundaries. Grupp (1994) argues that, in the future, technological innovation will not only be more dependent on basic research, but will also involve combinations of fields of technology, such as photonics, biosensors, and micro-technologies. In the future, continuing and extensive efforts will be required to seek synergies and 'fusion' between technologies. The challenge of management will be to create the strategies and structures to encourage this cross-disciplinary and cross-technology synergy.

These changes in techno-economic paradigm and in the possibilities for technology fusion provide many challenges and opportunities for business. The capacities of IT to share information more quickly, and the growing extent of globalization, both features of the 'knowledge economy', will make the effective and efficient use of technological innovation even more important sources of competitive advantage in the future. Furthermore, the increasing knowledge content of value-creating activities (discussed in Chapter 2) will continue to blur the boundaries between manufacturing and services. So, for example, a key future development in biotechnology lies in robotics—machines for the manufacture of genetic materials; major competitive advantages in the auto industry will derive from the provision of advantageous financial and insurance services; and customized software production will adopt some of the practices of larger-scale operations management. This will reduce the advantages of large-scale, entrenched positions, so that the opportunities for new entrants will increase. Given the rapid changes in technology, the ability of firms to move quickly, to learn fast and have the capacity to reach the market first, will confer competitive advantages.

How will firms respond to these opportunities and threats? The following organizational developments can be envisaged using Kay's (1993) concept of three types of organizational *architecture*: internal, external (between the firm and its suppliers or customers), and network (between a group of collaborating firms).

Changes in internal organization will continue along the current development path with all components of the business (for example, R & D, operations and production, and marketing) being ever more closely integrated. Importantly, this will bring a greater customer focus to production, driving manufacturing businesses towards more flexibility in design and production, and services firms towards more efficient production methods and niche and novel delivery strategies. Companies will be primarily organized by *process* rather than *function*, with the emphasis on building creative, learning organizations around the principles of small, cohesive groups (which Fairtlough 1994 calls compartmentation). These processes will be made increasingly more efficient, with an emphasis on the elimination of wasted effort and resources ('lean thinking').

Changes in external links will continue along the current best practice trajectory. Companies' innovation activities will be completely integrated with those of their suppliers and customers. Both customers and suppliers will be actively involved in defining both R & D activities and the new product development process (in real time, by means of electronic communications, from the product's conception to its commercialization). Individual consumers will be involved in the design and development of an increasing number of products and services, ranging from houses and cars to pharmaceuticals and education. Relationships within these external links will be based on high degrees of mutual commitment and trust in order to improve information flows, share learning, and encourage incremental innovation. The cooperation and integration of firms will extend to the formulation of their strategies. In supply chains, or Japanese *keiretsu*, this integration will continue to extend globally, particularly in Asia and Latin America. The development of new forms of linkage such as 'vendor pyramids', or 'global commodity chains', illustrates that even in industries dominated by large global firms there will be innovation opportunities for small and medium companies to form cascading links, as seen with companies like Boeing.

Within networks, increasing R & D collaboration will occur amongst clubs of firms with common interests and R & D service suppliers will be increasingly integrated into network activities. This strategic integration will be driven by the increasing cost and scope of innovation, and by the importance of time, or speed-to-market. Networking activity will see the development of new ways of operating, particularly in the management of relationships in which firms simultaneously cooperate and compete. Firms will become much more adept at integrating the insights and realizing the opportunities provided by public-sector research organizations, and research organizations will be highly skilled at proactively articulating the innovation needs of firms and finding solutions to them. Effective networks will be those that can integrate the activities of numerous constituents in order to deal with the complexity of technological systems.

These developments in networks are sympathetic to the recent analytic work on systems of innovation, discussed in Chapter 2. In future, the strength of national or regional innovation systems, and the competitiveness of individual firms, will lie not only in the strengths of the constituent elements—firms, research institutions, sources of finance—but more than ever in the linkages between them and the quality of their relationships.

Firms will also respond to the future challenges of technology-based competition by creating better technology strategies. Technology strategy will be formulated so as to improve MTI, and it will become the key means of identifying the aims and activities of technology-based firms. Importantly, technology strategy will improve the efficiency of technological investments and thereby allow firms to gain greater returns from a budgetary area that has been, and will remain, under strain. Few companies can maintain levels of investment in R & D at over 10 per cent of sales for any length of time without very substantial returns.

Companies will be skilled at selecting and developing technologies that satisfy and extend the needs of customers. Customers will increasingly be directly engaged with technology suppliers through the medium of e-commerce. Technology strategy will be formulated on the basis of extensive market intelligence, wide comprehension of the developments at the technological frontiers within internal and external research units, and competencies based on broad innovative capabilities and excellent technological and personnel resources. Technology strategy will be formulated on the basis of comprehensive understanding of the regulatory environment, and the articulation of that strategy will be instrumental in advising the development of those regulations and policy supports from government.

The promotion of internal technology transfer, the search for internal synergies across businesses, and the undertaking of more speculative R & D projects will be championed by the technology strategy unit, whose director (a board member) will report directly to the CEO. This unit will alleviate some of the shortcomings of centralized and decentralized R & D organizational structures, and the problems encountered by the changes companies have made in the past between one form and another.

The technology director will have the responsibility of providing stakeholders with an annual technology audit outlining present and future activities, and his or her performance will be measured not only by financial criteria but also by the expanded opportunity set and options made available to the firm. Technology audits will be part of firms' annual assessment of their knowledge bases, and will focus on evaluating and assessing intellectual property, and the scientific and technological skills residing in its staff.

The innovative companies of the future will be, like Benetton, innovative across the board, not just in technology. They will have the organizational structures, sources of finance, and marketing channels to build on the basis of their technological advances. Like Benetton, firms will find it commercially advantageous to take environmental issues seriously. In response to increasingly stringent regulatory requirements in areas like emissions control, and to strong customer and investor demand, the environmental credentials of firms and their products will be an ever-more important aspect of their competitiveness.

Just as industrial organizations will have to change in the future, so too will financial systems need to adjust. There will be a need for novel sources of funds (both domestic and international) to finance growth opportunities in the newly emerging technologies, particularly for longer-term technological developments. Banks will increasingly lose their power as money brokers as ICT directly links up buyers and sellers. The balance between equity and loan capital may change, as greater information availability through the Internet allows investors to make better decisions about the balance of investments in stocks or bonds. There will be less passive investment by venture capitalists, as seen in the European model, with the trend towards the interventionist style of US venture capital. The future will see growth in the numbers and quality of business angels, which is perhaps more appropriate for the increased entrepreneurship and growth management needed in the fifth and sixth technological waves. As a result of increased and more readily accessible information, managers and investors will be able to make more ethical decisions about technological investments. Whether the investment is in a production plant in a low-wage country with primitive regulations, or it is funding research into genetically manipulated seeds, companies will be progressively open to scrutiny about their decisions. Ethical issues will increasingly impinge upon the choices managers and investors make about technological investments.

New sources of 'exit'—where investors can realize their investments—will be needed for individual and institutional investors in start-ups in order to reduce their risk (valuable lessons can be learned from current experiences in over-the-counter markets, the Third Market in the United Kingdom, and the technology-based markets, NASDAQ in the United States and JASDAQ in Japan). Increased calls for accountability of finance managers to their investors will lead to increasing investor involvement in decision-making. There will be a need for new accounting tools, particularly related to valuing R & D options and reporting environmental and ethical credentials, and which will assist firms to take a longer-term perspective on their investments.

The management guru Peter Drucker has said that firms need to become learning and teaching organizations. In the knowledge economy, learning is the most important productive activity. As we have seen throughout this book, learning is a complicated, multifaceted activity (which has implications for simplistic notions of building 'learning organizations'). In the future, firms will need to understand, appreciate, and attempt to manage learning as, in the past, they attempted to manage other productive activities. They will require recruitment and training policies that attract and retain staff dedicated to lifelong learning.

MANAGING THE NEW INNOVATION PROCESS

We have seen how the innovation process is changing. One major characteristic of the fifth-generation innovation process is its increased 'automation' (Rothwell 1994). Electronic databases, simulation modelling, expert systems, artificial intelligence, and virtual-reality prototyping will be essential tools of the innovation process. Companies will utilize linked computer-integrated manufacturing systems integrating all elements of design and manufacturing. Firms will have electronic access to information in ERP systems in order to control the entire production system from the provision of raw materials to the final sale of products. Local area networks will integrate activities within production sites; electronic data interchange will link the activities of those sites with suppliers and customers; the information superhighway will assist the process of information exchange between firms and across all the activities within a firm from R & D to marketing. Just as automation revolutionized mass production in factories, the automation of innovation has the potential to revolutionize knowledge production.

These electronic tools both assist with, and add to, the complexity of the process of innovation. This complexity occurs in a number of forms. First, the environments in which firms operate will become increasingly complex. Change is rapid and unpredictable and, as we saw in Chapter 7, often requires technological collaborations and alliances in order to gain some control over the pace and direction of technological change (essentially sharing control in order to retain it). The scale and complexity of scientific advice, for example, are growing rapidly: the estimated 40,000–50,000 scientific journals in print in the 1990s will multiply. Secondly, the internal organizational structures required to ensure effective communications between a broader range of technical specialists will also become more complex (Von Glinow and Mohrman 1990). Thirdly, complexity will manifest in the configuration of 'product', which, as in the case of biotechnology, involves ever closer integration of recent basic research, and in the form of complex products and systems (CoPS), which, as seen in Chapter 2, comprise an important new form of business. Electronic tools may assist in dealing with complexity through facilitating the exchange and integration of information, but will add to the complexity by multiplying the potential sources of information (and placing greater technical demands on the security of commercially sensitive information).

The use of the electronic toolkit in the innovation process raises questions of whether knowledge that was previously tacit can be codified. If it can, then there are implications for appropriability and the use of collaborations (codified knowledge is more easily protected, and can be transferred via markets or hierarchies rather than alliances). Examination of the nature of knowledge in Chapter 2, however, showed that it is different from 'information' (Johnston 1998). Whereas the new electronic tools can store, transfer, and process information, their ability to do this with knowledge is limited. Tacit knowledge will remain a major differentiating factor for a firm's competitiveness, with the implication that rewards derive from investments in people. Particularly important will be investments in staff, who, as a result of investments in R & D and technology, will improve their ability to generate and assimilate new knowledge—that is, to learn.

As ecologists have discovered with biomass, healthy systems require diversity. There are many advantages for businesses in learning from diversity. As 3M's staff vice-president puts it, 'having people in our technical community who have grown up in different cultures provides us with a greater chance of innovation and creativity'. This diversity, as we saw in Chapter 7, also applies to choice of collaborating partners. Leading firms will seek a small number of tension-ridden alliances within or outside their networks, with sources of potentially disruptive technology to encourage learning and the development of radical innovations. Because of the technology collaboration paradox—where there are advantages in collaborating with the sort of firm that is more difficult to work with—there are greater than ever incentives to learning from failing (Maidique and Zirger 1985).

MANAGING RELATIONSHIPS WITH GOVERNMENT

The activities of governments play an important role in MTI. The regulations they establish—for example, in drug development—strongly influence MTI practices. The quality of intellectual-property protection they provide stimulates investments in R & D and in overseas technology investments. The extent to which they include technological innovation criteria in their procurement and outsourcing decisions will affect the overall level of innovation in a country. Standards-creating bodies enable the creation of technical standards that are broadly conceived and inclusive, rather than *de facto* and exclusive. Governments have directly supported innovation through a wide array of tax breaks, incentives, and direct grants for R & D, NPD, and investments in production. The development of technology in the East Asian economies has strongly depended upon government support. Even under the non-interventionist Reagan administration in the United States, a whole range of policies was put in place to encourage and facilitate technological innovation (Branscomb and Keller 1998).

Government innovation policies are changing (Dodgson and Bessant 1996), and will continue to undergo major changes. In future, these policies will have an ever-increasing focus on networks that will become major policy vehicles. These networks will build on local, regional, and national specialisms, and will provide a powerful mechanism for dealing with uncertainty and the idiosyncratic diversity of firms' needs. Firms, like people, come in a variety of shapes and sizes and exhibit widely varying characteristics. Designing policies that fit all contingencies is clearly impossible, but the alternative that governments will pursue is to design flexible arrangements that can be modulated by a network that shapes and adapts it to suit the individual needs of its constituent members. For firms to gain from innovation policies in the future, they will have to adopt a networking perspective.

A feature of this effort to increase networking will be the aim to improve the links in R & D between the public and private sectors. In Australia, for example, the Commonwealth Science and Industry Research Organization (CSIRO), which used to be fully funded by government, in the 1990s had to seek 30 per cent of its income from external sources. This required considerable change in the organization towards a more commercial orientation and greater preparedness to work with business. There are numerous examples of the privatization and corporatization of public-sector research bodies, from Malaysia to the United Kingdom, and this trend will continue.

Another feature of the new approach to innovation policy is the encouragement of a new type of innovation consultant. These innovation advisers are experienced in business and technology management and their role can be likened to the oil in the innovation engine, or to bees cross-pollinating innovative ideas. Some innovation advisers will be 'boundary spanners'—bringing in technology and market knowledge and insights from outside the network (often from overseas). As the skills of MTI become more appreciated, and achieve the status of professional associations, there will be increased outsourcing of MTI to expert managers.

For firms to value the policy support from government, these policies must be capable of development and adaptation. As we saw in the composite case of the firm receiving assistance with its development of machine tools in Taiwan, continual learning and adaptation is required by the policy-makers in order to deal with rapidly changing technologies. Policy-makers need not only to be abreast of science and technology development, but also to consider their social implications: the use of genetically manipulated organisms in food provides an example.

MANAGING SCIENCE AND BASIC RESEARCH

One of the most important roles of government in relation to MTI, and, in the knowledge economy, in relation to its overarching objectives of fostering economic and social development, is the encouragement of basic research. Basic science, technology, and innovation are interconnected, but the relationship between them is complex. Basic research is important for MTI, and MTI practices can be applied to the management of basic research. We have seen how basic research provides the basis for an increasing range of technologies and products. The trend away from longer-term, more basic research by business will be reversed in the future, and the present trend amongst leading industrial nations to extend investments in basic research will be continued. As an increasing amount of empirical research, new theories and analyses, and simple common sense and experience show that future economic success depends upon knowledge, then, as one of the fundamental drivers of knowledge, basic research will receive greater attention and investment.

According to Gibbons *et al.* (1995) there are two modes of knowledge production: the traditional and the new (see Chapter 2). Gibbons *et al.* unfortunately spend little time examining the impact of the new mode of knowledge creation for the management of research. However, they argue that the 'massification' of knowledge—the expansion of the university system, the increasing production of knowledge in government agencies, firms, and consultancies—has broken the monopoly of the traditional universities and produced an increasing number of competent researchers working outside the old system. They argue further that

> the advantages that these new locations of knowledge production have over traditional universities are two-fold. First, they offer more effective managerial models; in them, unlike the old universities, strategic planning is not inhibited by collegial government, nor tough choices obfuscated by the need to secure consensus. Second, they promise greater flexibility of response to fast-changing intellectual and professional needs; they seem to belong to a forward-looking enterprise culture sceptical of the traditional demarcations, taxonomies, hierarchies that clutter the old academic culture. (Gibbons *et al.* 1995: 82)

Research management is essential in new, Mode 2 knowledge production. As this mode is 'context specific', strategy (as described in Chapter 2) is essential for gaining consensus and unity amongst the diverse groups defining that context and shaping and conducting the research. As much of this research is collaborative, management is necessary to coordinate diverse inputs and plan the distribution of rewards that may be forthcoming from the research (such as ownership of IPRs).

Management has a more limited role to play in traditional, Mode 1 knowledge production. However, MTI skills are obviously needed in projects on the scale of nuclear research institutes, such as CERN, or large-scale, international collaborations, such as the Human Genome Project, which require a great deal of coordination and organization of facilities, personnel, materials, and public relations. Furthermore these projects, and universities, and large public laboratories, will benefit from the new paradigm management practices of strategy formulation, project management, and selection that contribute so much to the private sector.

Management can also contribute to improving the relationships between Mode 1 and Mode 2. Gibbons *et al.* do not assess the relationships between the different modes, apart from acknowledging that Mode 2 builds upon Mode 1. Management can play a role in encouraging the flow of personnel and their ideas between Mode 1 and Mode 2. The management of technology transfer in the University of Cambridge experience is illustrative here. Since the 1970s, management has created the environment in which academics enjoy forming links with firms, and indeed are motivated to move into industry to 'see projects through' for reasons of intellectual curiosity and feelings of 'ownership' of knowledge. As they enjoy security of tenure, they have the opportunity to return to traditional research, ensuring the continuing cross-fertilization of Mode 1 and Mode 2. Cambridge now has the largest concentration of high-tech industry in Europe. The cross-fertilization of the different forms of knowledge production will provide a major management challenge for the future.

It is a concern for the future that governments might concentrate more on funding Mode 2, rather than Mode 1, research because it is more manageable. Mode 1 research is essential, and it underpins Mode 2. For governments concerned with accountability, 'deliverables', and returns within the electoral cycle, the danger exists that traditional, disciplinary research will be ignored.

MANAGING GLOBAL SCIENCE AND TECHNOLOGY

Many of the major future challenges facing MTI are in its global management. These have been discussed in Chapter 3 and include issues of organizing international R & D and determining the extent to which the creation and production of product families are globalized. Another major management problem, and the one addressed here, is the question of which countries are the most important contributors and to beneficiaries from globalization.

The OECD refers to the growing phenomenon of *techno-globalism* and the need for considering the 'new rules of the game' that this implies. While technology is increasingly global, its ownership is becoming more unequal, with the technologically rich nations getting richer, and the poorer nations poorer (Whiston 1994). Indeed, even within the OECD, countries' technological strengths are becoming more differentiated (as measured by business commitment to R & D) (Patel and Pavitt 1994). As we have seen, more than 90 per cent of international strategic alliances in technology occur within and between the triad of the United States, Japan, and Europe (Dodgson 1993*a*). Participation in these global technology networks is hard earned. The challenge of those countries outside the triad is to be sufficiently technologically advanced and prepared to join these international networks. Superb science and distinctive technology are the only invitation to join the international networks. In the immediate future, however, the triad nations will continue to predominate global technology.

In any future for globalized technology, the countries of Asia will play a major role. We have seen the enthusiasm for (and difficulties of) doing business in China. As China develops its market economy, and as its thousands of research institutes, and many tens of thousands of talented researchers, develop a market-oriented approach, it will become a world technological powerhouse. It may, in the process, develop a Chinese style of innovation (which could, for example, involve overseas Chinese business networks, more intuitive sources of financial decision-making, and, depending upon political developments, close integration with the state).

The Japanese economy will continue to adapt and learn in the same way as it has in the past (Fruin 1992). It will invest more in basic science and improve the capacities of its universities both to do more basic research and to work more effectively with industry. As in the past, Japanese firms will continue paradoxically to be bastions of organizational conservatism and sources of organizational innovations. Their organizational innovations, which facilitate knowledge development and transfer, and technological fusion, will be copied overseas.

These alternative types of capitalism, and others found around the world, will continue to clash with the American style of capitalism (Gray 1998). It is to be hoped, for the reasons of diversity discussed earlier, and for social-equity reasons, that Gray's view about the continued ascendancy of entirely free market-based capitalism is unduly pessimistic.

Alternatives could continue and consolidate in Asia. The crisis that confronted some Asian countries at the end of the 1990s will be seen to have had a beneficial effect for their business and technological development. The radical changes occurring in Asian financial systems as a result of the crisis, with greater transparency and accountability, will lead to more efficient allocation of financial resources for technology projects. Government and business in these countries will see their views of science, technology, and innovation as being fundamental for future growth reinforced. The views held throughout Asian societies, which, based on past experience, strongly believe in the importance of technological innovation for economic and social progress, will also have been strengthened.

Small countries on the periphery of global technology development, such as Australia and Israel, will increasingly concentrate on specialist areas of science and technology. In Australia this will occur in areas such as earth sciences in the resources area, and in niche areas of medical and environmental science and technology. In Israel it will occur in the military and some electronics fields. India will continue to strangle the brilliance of its scientists and engineers with pathological degrees of bureaucracy and poor government. Countries in Latin America will continue to develop their particular advantages such as cheap labour (Mexico), large markets and some strong domestic firms (Brazil), and specialist expertise (forestry in Chile, and mining in Peru). The countries of the old Soviet bloc will build their technological strengths around their mineral resources, and the residual scientific expertise of its previously well-trained scientists. Northern and southern Africa will become the venue for increasing direct foreign investment in manufacturing plants. The demand for MTI in the countries of Central and Western Africa outside the petroleum and minerals sectors will, unfortunately, remain small.

All the challenges outlined in this chapter add to the difficulties of MTI described in the earlier chapters. MTI is one of the most formidable tasks confronting managers. As we have seen, it is also one of the most important. Technological innovation is a primary driver of competitiveness in the knowledge economy. In the twenty-first-century economy the performance of countries, firms, groups, and individuals critically depends upon the capacity to develop and effectively to use technological innovations of one sort or another. This book has examined the range of activities that managers need to master, from managing R & D and production and operations, to considerations of learning and trust. The argument is developed throughout the book that these issues have important *organizational* and *strategic* dimensions. The implications for practising, researching, and teaching MTI are clear: the management of any of the issues discussed here cannot be considered in isolation. Fundamental to MTI, for example, are questions of organizational integration—both internal and external—and the strategic combination of environmental assessment and definition of technological competencies. The book also shows the importance of the *international* dimension. Firms' activities are strongly influenced by global developments, and no country holds a monopoly on best practice in MTI. Future success will depend upon the capacity to address MTI with a global perspective, and to learn about MTI from wherever it is practised most efficiently and effectively around the world.

REFERENCES

Abernathy, W., and Clark, K. (1985), 'Innovation: Mapping the Winds of Creative Destruction', *Research Policy*, 14: 3–22.

—— and Utterback, J. (1978), 'Patterns of Industrial Innovation', *Technology Review* 80/7: 40–7.

Allen, T. (1977), *Managing the Flow of Technology: Technology Transfer and the Dissemination of Technological Information within the R & D Organization* (Cambridge, Mass.: MIT Press).

Ansoff, I. (1968), *Corporate Strategy* (Harmondsworth: Penguin).

Aoki, M. (1988), *Information, Incentives and Bargaining in the Japanese Economy* (Cambridge: Cambridge University Press).

—— and Rosenberg, N. (1987), 'The Japanese Firm as an Innovating Institution', paper presented at the Institutions in a New Democratic Society Conference, 15–17 Sept.

Archibugi, D., and Michie, J. (1995), 'The Globalization of Technology: A New Taxonomy', *Cambridge Journal of Economics*, 19: 121–40.

Argyris, C., and Schon, D. (1978), *Organizational Learning* (London: Addison Wesley).

Arnold, E., Guy, K., and Dodgson, M. (1992), *Linking for Success: Making the Most of Collaborative R & D* (London: National Economic Development Office/Institution of Electrical Engineers).

Arrow, K. (1962), 'The Economic Implications of Learning by Doing', *Review of Economic Studies*, 29/2: 155–73.

—— (1975), 'Gifts and Exchanges', in E. Phelps (ed.), *Altruism, Morality and Economic Theory* (New York: Russell Sage).

Arthur, B. (1990), 'Positive Feedbacks in the Economy', *Scientific American* (Feb.), 80–5.

AT Kearney (1989), *Computer-Integrated Manufacturing—Competitive Advantage or Technological Dead-End?* (London: AT Kearney Consultants).

Augsdorfer, P. (1994), 'Taxonomy of Management Attitudes towards Bootlegging and Uncertainty', in R. Oakey (ed.), *High Technology Small Firms in the 1990s* (London: Paul Chapman).

Axelrod, R. (1984), *The Evolution of Cooperation* (Harmondsworth: Penguin).

Baba, Y. (1989), 'Characteristics of Innovating Japanese Firms-Reverse Product Cycles', in M. Dodgson (ed.), *Technology Strategy and the Firm: Management and Public Policy* (Harlow: Longman).

Baird, I., Lyles, M., and Wharton, R. (1990), 'Attitudinal Differences between American and Chinese Managers Regarding Joint Venture Management', *Management International Review*, 30 (Special Issue), 53–68.

Baker, K. (1997), 'Leading the Globalization of the Product Development Process', paper presented at the Global Automotive Management Council, Erice, Italy, 28 July.

Bandura, A. (1977), *Social Learning Theory* (Boston: MIT Press).

Barnet, R., and Cavanagh, J. (1994), *Global Dreams* (New York: Simon & Schuster).

Barras, R. (1986), 'Towards a Theory of Innovation in Services', *Research Policy*, 15/4: 161–73.

BASF (1997), 'BASF Takes a Stand—Globalization', BASF News Release (Apr.).

Beatty, C., and Gordon, J. (1991), 'Preaching the Gospel: The Evangelists of New Technology', *California Management Review* (Spring).

Behrman, J., and Fischer, W. (1980), *Overseas Activities of Transnational Companies* (Cambridge, Mass.: Oelgeschlager, Gunn and Hain).

Belussi, F. (1989), 'Benetton: A Case Study of Corporate Strategy for Innovation in Traditional Sectors', in M. Dodgson (ed.), *Technology Strategy and the Firm: Management and Public Policy* (Harlow: Longman).

Bessant, J. (1991), *Managing Advanced Manufacturing Technology: The Challenge of the Fifth Wave* (Oxford: Blackwell).

—— (1993), 'The Lessons of Failure: Learning to Manage New Technology', *International Journal of Technology Management*, 8/3–5: 197–215.

—— (1998), 'Developing Continuous Improvement Capability', *International Journal of Innovation Management*, 2/4: 409–29.

—— and Buckingham, J. (1993), 'Organizational Learning for Effective Use of CAPM', *British Journal of Management*, 4/4: 219–34.

Best, M. (1990), *The New Competition* (London: Polity Press).

Betz, F. (1988), Managing Technological Innovation (New York: Wiley).

Bianchi, P., and Bellini, N. (1991), 'Public Policies for Local Networks of Innovators', *Research Policy*, 20: 487–97.

Bidault, F. (1989), *Technology Pricing* (New York: St. Martin's Press).

Birley, S. (1982), 'Corporate Strategy in the Small Firm', *Journal of General Management*, 8/2: 82–6.

Bitondo, D., and Frohman, A. (1981), 'Linking Technological and Business Planning', *Research Management*, 24/6: 19–23.

Blakeney, M. (1989), *Legal Aspects of the Transfer of Technology to Developing Countries* (Oxford: ESC Publishing).

Bloom, M. (1992), *Technological Change in the Korean Electronics Industry* (France: OECD).

Bourke, P., and Butler, L. (1995), *International Links in Higher Education Research* (Canberra: National Board of Employment, Education and Training).

Bowinder, B., and Miyake, T. (1993), 'Japanese Innovations in Advanced Technologies: An Analysis of Functional Integration', *International Journal of Technology Management*, 8/1–2: 135–56.

Branscomb, L., and Keller, J. (1998) (eds.), *Investing in Innovation* (Cambridge, Mass.: MIT Press).

Brown, S. (1996), *Strategic Manufacturing for Competitive Advantage* (London: Prentice Hall).

—— and Eisenhardt, K. (1995), 'Product Development: Past Research, Present Findings and Future Directions', *Academy of Management Review*, 20/2: 343–78.

Buckley, P., and Casson, M. (1988), 'A Theory of Cooperation in International Business', in F. Contractor and P. Lorange (eds.), *Cooperative Strategies in International Business* (Lexington, Mass.: Lexington Books).

Burgelman, R., Maidique, M., and Wheelwright, S. (1996), *Strategic Management of Technology and Innovation* (Chicago: Irwin).

Burns, T., and Stalker, G. (1961), *The Management of Innovation* (London: Tavistock).

Cantwell, J. (1998), 'Innovation as the Principal Source of Growth in the Global Economy', in D. Archibugi, J. Howells, and J. Michie (eds.), *Innovation in the Global Economy* (Cambridge: Cambridge University Press).

—— and Andersen, B. (1996), 'A Statistical Analysis of Corporate Technological Leadership Historically', *Economics of Innovation and New Technology*, 4 : 211–34.

Carlsson, B. (1994), 'Technological Systems and Economic Performance', in M. Dodgson and R. Rothwell. (eds.), *The Handbook of Industrial Innovation* (Aldershot: Edward Elgar).

—— and Stankiewicz, R. (1991), 'On the Nature, Function, and Composition of Technological Systems', *Journal of Evolutionary Economics*, 1/2: 93–118.

Carnegie Bosch Institute (1995), *Organization of R & D to Support Global and Regional Markets* (Pittsburgh: Carnegie Bosch Institute).

Carrubba, F. (1993), 'An International Approach to Research—Philips Electronics NV', paper presented at the Strategies for Effective Research Conference, Royal Academy of Engineering, London, 10 Mar.

Carter, C., and Williams, B. (1957), *Industry and Technical Progress: Factors Governing the Speed of Application of Science* (London: Oxford University Press).

Casson, M. (1991), *Global Research Strategy and International Competitiveness* (Oxford: Blackwell).

Castells, M. (1996), *The Rise of the Network Society* (Oxford: Blackwell).

Chadwick, R. (1996), *Practising Balance: Integrating Best Financial Practice into your Business* (Melbourne: Australian Manufacturing Council).

Chan, T. (1997), 'Global Technology Integration—A Cornerstone to Globalizing Business Operations', paper presented at the Technology-Application-Market International Forum, Tsinghua University, Beijing, Nov.

Chandler, A. (1966), *Strategy and Structure* (Cambridge, Mass.: MIT Press).

Chen, S.-H. (1996), 'Decision-Making in Research and Development Collaboration', *Research Policy*, 26/1: 121–35.

Christensen, C. (1997), *The Innovator's Dilemma* (Cambridge, Mass.: Harvard Business School Press).

Ciborra, C. (1991), 'Alliances as Learning Experiences: Cooperation, Competition and Change in High-Tech Industries', in L. Mytelka (ed.), *Strategic Partnerships and the World Economy* (London: Pinter Publishers).

—— and Andreu, R. (1998), 'Organizational Learning and Core Capabilities Development: The Role of IT', in R. Galliers and W. Baet (eds.), *Information Technology and Organizational Transformation* (Chichester: Wiley).

Clark, K. (1989), 'Project Scope and Project Performance: The Effects of Parts Strategy and Supplier Involvement on Product Development', *Management Science*, 35/10: 1247–63.

—— and Fujimoto, T. (1991), *Product Development Performance* (Boston: Harvard Business School Press).

—— and Wheelwright, S. (1993), *Managing New Product and Process Development* (Cambridge, Mass.: Free Press).

—— Hayes, R., and Lorenz, C. (1987), *The Uneasy Alliance: Managing the Technology-Productivity Dilemma* (Boston: Harvard Business School Press).

Coe, D., and Helpman, E. (1993), *International R & D Spillovers* (Washington: International Monetary Fund).

Cohen, W., and Levinthal, D. (1990), 'Absorptive Capacity: A New Perspective on Learning and Innovation', *Administrative Science Quarterly*, 35: 128–52.

Cole, R. (1995) (ed.), *The Death and Life of the American Quality Movement* (New York: Oxford University Press).

Collins, J., and Porras, J. (1994), *Built to Last: Successful Habits of Visionary Companies* (New York: Harper Collins).

Contractor, F., and Lorange, P. (1988), 'Why Should Firms Cooperate? The Strategy and Economics Basis for Cooperative Ventures', in F. Contractor and P. Lorange (eds.), *Cooperative Strategies in International Business* (Lexington, Mass.: Lexington Books).

Cooke, P., and Morgan, K. (1991), 'The Network Paradigm: New Departures in Corporate and Regional Development' working paper (Cardiff: University of Wales at Cardiff).

—— —— (1994), 'The Creative Milieu: A Regional Perspective on Innovation', in M. Dodgson and R. Rothwell (eds.), *The Handbook of Industrial Innovation* (Aldershot: Edward Elgar).

Coombs, R. (1994), 'Technology and Business Strategy', in M. Dodgson and R. Rothwell (eds.), *The Handbook of Industrial Innovation* (Aldershot: Edward Elgar).

—— and Richards, A. (1991), 'Technologies, Products and Firms' Strategies', *Technology Analysis and Strategic Management*, 3: 77–86.

Cooper, R. (1980), 'Project NewProd: Factors in New Product Success', *European Journal of Marketing*, 14/5–6: 277–92.

—— (1993), *Winning at New Products* (Reading, Mass.: Addison-Wesley).

—— (1994), 'Third Generation New Product Processes', *Journal of Product Innovation Management*, 11: 3–14.

—— and Kleinschmidt, E. (1996), 'Winning Businesses in Product Development: Critical Success Factors', *Research-Technology Management*, 39/4: 198–229.

Cornwell, B. (1998), 'Quicklook Commercialization Assessments', *R & D Enterprise—Asia Pacific*, 1/1: 7–9.

Cringely, R. (1996), *Accidental Empires* (Harmondsworth: Penguin).

Cusumano, M., and Selby, R. (1995), *Microsoft Secrets* (New York: Free Press).

Dalton, D., and Serapio, M. (1995), *Globalizing Industrial Research and Development* (Washington: US Department of Commerce).

David, P. (1986), 'Clio and the Economics of QWERTY', in W. Parker (ed.), *Economic History and the Modern Economist* (Oxford: Blackwell).

Davies, A. (1997), 'Life Cycle of a Complex Product System', *International Journal of Innovation*, 1/3: 229–56.

De Meyer, A. (1992), 'Internationalisation of R & D', paper presented at the Third International Conference in Science and Technology Policy Research, 9–11 Mar.

—— (1993), 'Management of an International Network of Industrial R & D Laboratories', *R & D Management*, 23/2.

DeBresson, C., and Amesse, F. (1991), 'Networks of Innovators: A Review and Introduction to the Issue', *Research Policy*, 20: 363–79.

Dertouzos, M., Lester, R., and Solow, R. (1989), *Made in America: Regaining the Productive Edge* (New York: HarperPerennial).

Di Maggio, P., and Powell, W. (1983), 'The Iron Cage Revisited: Institutional Isomorphism and Collective Rationality in Organizational Fields', *American Sociological Review*, 48: 147–60.

Dixit, A., and Pindyck, R. (1995), 'The Options Approach to Capital Investment', *Harvard Business Review* (May–June), 105–15.

Dodgson, M. (1985), *Advanced Manufacturing Technology in the Small Firm-Variation in Use and Lessons for the Flexible Organisation of Work* (London: Technical Change Centre).

—— (1989) (ed.), *Technology Strategy and the Firm: Management and Public Policy* (Harlow: Longman).

—— (1990*a*), *Celltech: The First Ten Years of a Biotechnology Company* (SPRU: University of Sussex).

—— (1990*b*), 'The Shock of the New: The Formation of Celltech and the British Technology Transfer System', *Industry and Higher Education*, 4/2: 97–104.

—— (1991), *The Management of Technological Learning* (Berlin: De Gruyter).

—— (1993*a*), *Technological Collaboration in Industry: Strategy Policy and Internationalization in Innovation* (London, Routledge).

—— (1993*b*), 'Learning, Trust and Technological Collaboration'. *Human Relations*, 46/1: 77–95.

—— (1993*c*), 'Organizational Learning: A Review of Some Literatures', *Organization Studies*, 14/3: 375–94.

—— and Bessant, J. (1996), *Effective Innovation Policy: A New Approach* (London: International Thomson Business Press).

—— and Kim, Y. (1997), 'Learning to Innovate: Korean Style—the Case of Samsung', *International Journal of Innovation Management*, 1/1: 53–67.

—— and Rothwell, R. (1991), 'Technology Strategy in Small Firms', *Journal of General Management*, 17/1: 45–55.

——, Sako, M., and Sapsed, J. (1995), 'Achieving Complementarities of Size Advantages in New Product Development: The Case of Multimedia in Japan', *International Journal of Innovation Management*, Special Edition in Emerging Technological Frontiers to International Competition, 183–205.

Dore, R., and Sako, M. (1998), *How the Japanese Learn to Work* (London: Routledge).

Dosi, G. (1982), 'Technological Paradigms and Technological Trajectories: A Suggested Interpretation of the Determinants and Directions of Technical Change', *Research Policy*, 2/3: 147–62.

—— (1988), 'Sources, Procedures, and Microeconomic Effects of Innovation', *Journal of Economic Literature*, 26: 1120–71.

Doz, Y. (1988), 'Technology Partnerships between Larger and Smaller Firms: Some Critical Issues', in F. Contractor and P. Lorange (eds.), *Cooperative Strategies in International Business* (Lexington, Mass.: Lexington Books).

—— and Schuen, A. (1988), 'From Intent to Outcome: A Process Framework for Partnerships', paper presented at the Prince Bertil Symposium—Corporate and Industry Strategies for Europe, Stockholm, 9–11 Nov.

Dreher, C. (1996), 'The Use of Modern Manufacturing Concepts by German Capital Equipment Producers', Lecture at University of Michigan Business School.

Dussauge, P., Hart, S., and Ramanantsoa, B. (1993), *Strategic Technology Management* (Chichester: Wiley).

Economic Planning Agency (1990), *Keizai Hakusho* (Tokyo: Okurasho).

Edquist, C. (1997), *Systems of Innovation* (London: Pinter).

EIRMA (1986): European Industrial Research Managers Association, *Developing R & D Strategies* (Paris: EIRMA).

Ettlie, J., Dreher, C., Kovacs, G., and Trygg, L. (1993), 'Cross-National Comparisons of Product Development in Manufacturing', *Journal of High Technology Management Research*, 4/2: 139–55.

European Commission (1998), *Internationalization of Research and Technology* (Brussels: European Technology Assessment Network).

Fagerberg, J. (1987), 'A Technology-Gap Approach to Why Growth Rates Differ', *Research Policy*, 16: 87–99.

Fairtlough, G. (1994*a*), *Creative Compartments: A Design for Future Organisation* (London: Adamantine Press).

Fairtlough, G. (1994*b*), 'Innovation and Organization', in M. Dodgson and R. Rothwell (eds.), *The Handbook of Industrial Innovation* (Aldershot: Edward Elgar).

Farris, G., DiTomaso, N., and Cordero, R. (1995), 'Over the Hill and Losing It? The Senior Scientist and Engineer', conference paper, Academy of Management, Vancouver, 6–9 Aug.

Fiol, C., and Lyles, M. (1985), 'Organisational Learning', *Academy of Management Review*, 10/4: 803–13.

Florida, R. (1994), 'Reinventing R & D: The Changing Face of Innovation in the American Economy', mimeo (Pittsburg: Carnegie Mellon University).

—— (1997), 'The Globalization of R & D: Results of a Survey of Foreign-Affiliated R & D Laboratories in the United States', *Research Policy*, 26: 85–103.

Fox, A. (1974), *Beyond Contract: Work Power and Trust Relations* (London: Faber and Faber).

Freeman, C. (1974), *The Economics of Industrial Innovation* (London: Pinter).

—— (1982), *The Economics of Industrial Innovation* (London: Pinter).

—— (1987), *Technology Policy and Economic Performance: Lessons from Japan* (London: Pinter).

—— (1991), 'Networks of Innovators: A Synthesis of Research Issues', *Research Policy*, 20: 499–514.

—— (1994), 'Innovation and Growth', in M. Dodgson and R. Rothwell (eds.), *The Handbook of Industrial Innovation* (Aldershot: Edward Elgar).

—— and Perez, C. (1988), 'Structural Crises of Adjustment: Business Cycles and Investment Behaviour', in G. Dosi, C. Freeman, R. Nelson, G. Silverberg, and L. Soete (eds.), *Technical Change and Economic Theory* (London: Pinter).

—— and Soete, L. (1997), *The Economics of Industrial Innovation* (London: Pinter).

Fruin, M. (1992), *The Japanese Enterprise System* (Oxford: Oxford University Press).

Fukuyama, F. (1995), *Trust: The Social Virtues and the Creation of Prosperity* (New York: Free Press).

Furukawa, K., Teramoto, Y. and Kanda, M. (1990), 'Network Organization for Inter-Firm R & D Activities: Experiences of Japanese Small Businesses', *International Journal of Technology Management*, 5/1: 27–40.

Fusfield, A. (1978), 'How to Put Technology into Corporate Planning', *Technology Review* (May), 51–5.

Gann, D. (1991), 'Technological Change and the Internationalisation of Construction in Europe', in C. Freeman, M. Sharp, and W. Walker (eds.), *Technology and the Future of Europe* (London: Pinter Publishers).

—— (1994), 'Innovation in the Construction Industry', in M. Dodgson and R. Rothwell (eds.), *The Handbook of Industrial Innovation* (Aldershot: Edward Elgar).

—— and Salter, A. (1998), 'Learning and Innovation Management in Project-Based, Service-Enhanced Firms', *International Journal of Innovation Management*, 2/4: 431–54.

Gardiner, P., and Rothwell, R. (1985), 'Tough Customers: Good Designs', *Design Studies*, 6/1: 7–17.

Garvin, D. (1992), *Operations Strategy, Text and Cases* (Englewood Cliffs, NJ: Prentice Hall).

—— (1993), 'The Learning Organization', *Harvard Business Review* (July–Aug.), 78–91.

Gerybadze, A., and Reger, G. (1997), 'Globalisation of R & D: Recent Changes in the Management of Innovation in Transnational Corporations, report (Stuttgart: Hohenheim University).

—— —— (1998), 'Managing Globally Distributed Competence Centers within Multinational Corporations: A Resource-Based View', in T. Scandura and M. Serapio (eds.), *Research in International Business and International Relations—Leadership and Innovation in Emerging Markets* (Stamford, Conn.: JAI Press).

—— —— (1999), 'Globalization of R & D: Recent Changes in the Management of Innovation in Transnational Corporations', *Research Policy*, 28/2–3: 251–74.

Gibbons, M., Limoges, C., Nowotny, H., Schwartzmann, S., Scott, P., and Trow, M. (1995), *The New Production of Knowledge* (London: Sage).

Goldberg, A., and. Shenhav, Y. (1984), 'R & D Career Paths: Their Relation to Work Goals and Productivity', *IEEE Transactions on Engineering Management*, EM-31: 111–17.

Goodman, R., and Lawless, M. (1994), *Technology and Strategy: Conceptual Models and Diagnostics* (New York: Oxford University Press).

Granstrand, O., and Sjolander, S. (1990), 'Managing Innovation in Multi-Technology Corporations', *Research Policy*, 19/1: 35–60.

—— Hakansson, L., and Sjolander, S. (1992), *Technology Management and International Business* (London: John Wiley).

Grant, R. (1991), *Contemporary Strategy Analysis* (Cambridge, Mass.: Blackwell).

Graves, A. (1994), 'Innovation in a Globalizing Industry: The Case of Automobiles', in M. Dodgson and R. Rothwell (eds.), *The Handbook of Industrial Innovation* (Aldershot: Edward Elgar).

Gray, J. (1998), *False Dawn: The Delusions of Global Capitalism* (London: Granta).

Griliches, Z. (1986), 'Productivity, R & D and Basic Research at the Firm Level in the 1970s', *American Economic Review*, 76/1: 141–54.

Grupp, H. (1992), *Dynamics of Science-Based Innovation* (Heidelberg: Springer).

—— (1994), 'The Dynamics of Science-Based Innovation Reconsidered', in O. Granstrand (ed.), *Economics of Technology* (Amsterdam: Elsevier).

—— (1996), 'Spillover Effects and the Science Base of Innovation Reconsidered: An Empirical Macro-Economic Approach', *Journal of Evolutionary Economics*, 6: 175–97.

Hagedoorn, J. (1997), 'International Technological Collaboration and Firm Dynamics: Implications for NIEs', paper presented at the STEPI Conference on Innovation and Competitiveness in Newly Industrializing Economies, Seoul, 26–7 May.

Hakansson, H. (1982), *International Marketing and Purchasing of Industrial Goods: An Interaction Approach* (Chichester, Wiley).

—— and Johanson, J. (1988), 'Formal and Informal Cooperation Strategies in International Industrial Networks', in F. Contractor and P. Lorange (eds.), *Cooperative Strategies in International Business* (Lexington, Mass.: Lexington Books).

—— and Snehota, I. (1995), *Developing Relationships in Business Networks* (London: Routledge).

Halberstam, D. (1987), *The Reckoning* (London, Bloomsbury).

Halliday, R., Walker, S., and Lumley, C. (1992), 'R & D Philosophy and Management in the World's Leading Pharmaceutical Firms', *Journal of Pharmaceutical Medicine*, 2: 139–54.

—— Drasdo, A., Lumley, C., and Walker, S. (1997), 'The Allocation of Resources for R & D in the World's Leading Pharmaceutical Companies', *R & D Management*, 27/1: 63–77.

Harding, R. (1995), *Technology and Human Resources in their National Context* (Aldershot, Avebury).

Harrigan, K. (1986), *Managing for Joint Venture Success* (Lexington, Mass.: Lexington Books).

Harris, J., Shaw, R., and Sommers, W. (1983), 'The Strategic Management of Technology', *Planning Review* (Jan.), 28–35.

Hayes, R., and Pisano, G. (1994), 'Beyond World-Class: The New Manufacturing Strategy', *Harvard Business Review* (Jan.–Feb.): 77–86.

—— and Wheelwright, S. (1984), *Restoring our Competitive Edge: Competing through Manufacturing* (New York: Wiley).

—— Wheelwright, S., and Clark, K. (1988), *Dynamic Manufacturing: Creating the Learning Organization* (New York: Free Press).

Hedberg, B. (1981), 'How Organizations Learn and Unlearn', in P. Nystrom and W. Starbuck (eds.), *Handbook of Organizational Design*, i. (Oxford: Oxford University Press).

Henderson, R. (1994), 'Managing Innovation in the Information Age', *Harvard Business Review*, 72/1: 100–5.

—— and Clark, K. (1990), 'Architectural Innovation', *Administrative Science Quarterly*, 35: 9–30.

Hicks, D. (1995), 'Published Papers, Tacit Competencies and Corporate Management of the Public/Private Character of Knowledge', *Industrial and Corporate Change*, 4/2: 401–24.

Hill, T. (1985), *Manufacturing Strategy* (Basingstoke, Macmillan).

Hirst, P., and Zeitlin, J. (1989), *Reversing Industrial Decline* (London: Berg).

Hobday, M. (1994), 'Innovation in East Asia: Diversity and Development', in M. Dodgson and R. Rothwell (eds.), *The Handbook of Industrial Innovation* (Aldershot: Edward Elgar).

—— (1995), *Innovation in East Asia* (Aldershot: Edward Elgar).

—— (1998), 'Product Complexity, Innovation and Industrial Organization', *Research Policy*, 26/6: 689–710.

Hofstede, G. (1980), *Cultural Consequences: International Differences in Work-Related Values* (Newbury Park, Calif.: Sage).

Howard, R. (1993), *The Learning Imperative* (Boston, Mass.: Harvard Business School Press).

Howells, J. (1990), 'The Location and Organization of Research and Development: New Horizons', *Research Policy*, 19: 133–46.

—— and Wood, M. (1993), *The Globalisation of Production and Technology* (London: Belhaven Press).

Hrebreniak, L. (1978), *Complex Organizations* (St Paul, Minn.: West Publishing).

Hull, F., and Azumi, K. (1991), 'Invention Rates and R & D in Japanese Manufacturing Organizations', *Journal of Engineering and Technology Management*, 8: 37–66.

Iansiti, M. (1993), 'Real-World R & D: Jumping the Product Generation Gap', *Harvard Business Review* (May–June), 138–46.

Imai, K., Nonaka, I., and Takeuchi, H. (1988), 'Managing the New Product Development Process: How Japanese Companies Learn and Unlearn', in M. Tushman and W. Moore (eds.), *Readings in the Management of Innovation* (New York: Harper Business).

Irvine, J. (1988), *Evaluating Applied Research: Lessons from Japan* (London: Pinter).

Itoh, M., and Urata, S. (1994), *Small and Medium Enterprise Support Policies in Japan* (Tokeyo: University of Tokyo).

Jarillo, J. (1988), 'On Strategic Networks', *Strategic Management Journal*, 19: 31–41.

Johnston, R. (1998), *The Changing Nature and Forms of Knowledge: A Review* (Canberra, Australia: Department of Education, Employment, Training, and Youth Affairs (DEETYA)).

—— and Blumentritt, R. (1988), 'Knowledge Moves to Centre Stage', *Science Communication*, 20/1: 99–105.

Jolly, V. (1997), *Commercializing New Technologies: Getting from Mind to Market* (Boston: Harvard Business School Press).

Jones, O., and Smith, D. (1997), 'Strategic Technology Management in a Mid-Corporate Firm: The Case of Otter Controls', *Journal of Management Studies*, 34/4: 511–36.

Judge, W., Fryxell, G., and Dooley, R. (1997), 'The New Task of R & D Management: Creating Goal-Directed Communities for Innovation', *California Management Review*, 39/3: 72–85.

Kamath, R., and Liker, J. (1990), 'Supplier Dependence and Innovation: A Contingency Model of Suppliers' Innovative Activities', *Journal of Engineering and Technology Management*, 7: 11–127.

Kamien, M., and Schwartz, N. (1975), 'Market Structure and Innovation', *Journal of Economic Literature*, 23/1: 1–37.

Kaplinsky, R. (1984), *Automation* (Harlow: Longman).

Katz, R., and Allen, T. (1982), 'Investigating the Not Invented Here (NIH) Syndrome: A Look at the Performance, Tenure, and Communication Patterns of 50 R & D Project Groups', *R & D Management*, 12/1: 7–19.

Kay, J. (1993), *Foundations of Corporate Success* (Oxford: Oxford University Press).

Kenney, M., and Florida, R. (1993), *Beyond Mass Production* (New York: Oxford University Press).

—— —— (1994), 'The Organization and Geography of Japanese R & D: Results from a Survey of Japanese Electronics and Biotechnology Firms', *Research Policy*, 23: 305–23.

Kim, L. (1993), *National System of Industrial Innovation: Dynamics of Capability Building in Korea* (New York: Oxford University Press).

—— (1997), *Imitation to Innovation: The Dynamics of Korea's Technological Learning* (Boston, Mass.: Harvard Business School Press).

Kline, S., and Rosenberg, N. (1986), 'An Overview of Innovation', in R. Landau and N. Rosenberg (eds.), *The Positive Sum Strategy* (Washington: National Academy Press).

Kodama, F. (1987), 'Japanese Innovation in Mechatronics Technology', *Science and Public Policy*, 12/1: 291–6.

—— (1995), *Emerging Patterns of Innovation* (Boston, Mass.: Harvard Business School Press).

Kogut, B. (1988*a*), 'Joint Ventures: Theoretical and Empirical Perspectives', *Strategic Management Journal*, 9: 312–32.

—— (1988*b*), 'A Study of the Life Cycle of Joint Ventures', in F. Contractor and P. Lorange (eds.), *Cooperative Strategies in International Business* (Lexington, Mass.: Lexington Books).

—— and Zander, U. (1993), 'Knowledge of the Firm and the Evolutionary Theory of the Multinational Firm', *Journal of International Business Studies* (Fourth Quarter), 625–45.

Kotha, S. (1995), 'Mass Customization: Implementing the Emerging Paradigm for Competitive Advantage', *Strategic Management Journal*, 16: 21–42.

Kuemmerle, W. (1997), 'Building Effective R & D Capabilities Abroad', *Harvard Business Review* (Mar.– Apr.), 61–70.

Lamming, R. (1992), 'Supplier strategies in the Automotive Components Industry: Development towards Lean Production, Ph.D. thesis, University of Sussex, Falmer.

Lauglaug, A. (1987), 'A Framework for the Strategic Management of Future Tyre Technology', *Long Range Planning*, 20/5: 21–41.

Leonard-Barton, D. (1992), 'The Factory as a Learning Laboratory', *Sloan Management Review*, 34/1: 23–38.

—— (1995), *Wellsprings of Knowledge* (Boston, Mass.: Harvard Business School Press).

Lester, R. (1998), *The Productive Edge: How US Industries are Pointing the Way to a New Era of Economic Growth* (New York: W. W. Norton).

Levinthal, D., and March, J. (1981), 'A Model of Adaptive Organizational Search', *Journal of Economic Behaviour and Organization*, 2: 307–33.

Levitt, B., and March, J. (1988), 'Organisational Learning', *Annual Review of Sociology*, 14: 319–40.

Littler, D. (1994), 'Marketing and Innovation', in M. Dodgson and R. Rothwell (eds.), *The Handbook of Industrial Innovation* (Aldershot: Edward Elgar).

Loveridge, R., and Pitt, M. (1990) (eds.), *The Strategic Management of Technological Innovation* (Chichester: Wiley).

Lundvall, B.-A. (1988), 'Innovation as an Interactive Process: from User–Producer Interaction to the National System of Innovation', in G. Dosi, C. Freeman, R. Nelson, G. Silverberg, and L. Soete (eds.), *Technical Change and Economic Theory* (London: Pinter Publishers).

—— (1992) (ed.), *National Systems of Innovation* (London: Pinter).

Lutz, R. (1994), 'Implementing Technological Change with Cross-Functional Teams', *Research-Technology Management*, 14: 14–18.

Lyles, M. (1988), 'Learning among Joint Venture-Sophisticated Firms', in F. Contractor and P. Lorange (eds.), *Cooperative Strategies in International Business* (Lexington, Mass.: Lexington Books).

Macauley, S. (1963), 'Non-Contractual Relations in Business: A Preliminary Study', *American Sociological Review*, 28/1: 55–66.

Macdonald, S., and Williams, C. (1994), 'The Survival of the Gatekeeper', *Research Policy*, 23: 123–32.

McGrath, M., and Romeri, M. (1994), 'The R & D Effectiveness Index: A Metric for New Product Development Performance', *Journal of Product Innovation Management*, 11: 213–20.

Maidique, M., and Zirger, B. (1985), 'The New Product Learning Cycle', *Research Policy*, 14: 299–313.

Malerba, F. (1992), 'Learning by Firms and Incremental Technical Change', *Economic Journal*, 102: 845–59.

Manley, K. (1994), 'Factors Leading to Offshore Manufacture of Australian Inventions', Ph.D. thesis, Murdoch University, Perth.

Mansell, R., and Wehn, U. (1998), *Knowledge Societies: Information Technology for Sustainable Development* (Oxford: Oxford University Press).

Mansfield, E. (1988*a*), 'Industrial Innovation in Japan and the United States', *Science*, 241: 1769–74.

—— (1988*b*), 'Industrial R & D in Japan and the United States', *American Economic Review*, 78: 223–8.

—— (1998), 'Academic Research and Industrial Innovation: An Update of Empirical Findings', *Research Policy*, 26: 773–6.

—— Rapoport, J., Romeo, A., Wagner, S., and Beardsley, G. (1977), 'Social and Private Rates of Return from Industrial Innovations', *Quarterly Journal of Economics*, 91/2: 221–40.

Marceau, J. (1992) (ed.), *Reworking the World: Organisations, Technologies and Cultures in Comparative Perspective* (Berlin: De Gruyter).

—— (1994), 'Clusters, Chains and Complexes: Different Approaches to Innovation with a Public Policy Perspective', in M. Dodgson and R. Rothwell (eds.), *The Handbook of Industrial Innovation* (Aldershot: Edward Elgar).

March, J. (1991), 'Exploration and Exploitation in Organizational Learning', *Organization Science*, 2/1: 71–87.

—— Sproull, L., and Tamuz, M. (1991), 'Learning from Samples of One or Fewer', *Organization Science*, 2/1: 1–13.

Marengo, L. (1992), 'Coordination and Organizational Learning in the Firm', *Journal of Evolutionary Economics*, 2: 313–26.

Martin, B., and Johnston, R. (1998), *Technology Foresight for Wiring up the National Innovation System: Experiences in Britain, Australia and New Zealand* (Brighton: Science Policy Research Unit).

—— and Salter, A. (1996), *The Relationship Between Publicly Funded Basic Research and Economic Performance* (Brighton: Science Policy Research Unit).

Maschlup, F. (1982), *Knowledge: Its Creation, Distribution and Economic Significance* (Princeton: Princeton University Press).

Mathews, J. (1996), 'High Technology Industrialisation in East Asia', *Journal of Industry Studies*, 3/2: 1–69.

Meyer, C. (1993), *Fast Cycle Time* (New York: Free Press).

Meyer, M., and Lehnerd, A. (1997), *The Power of Product Platforms* (New York: Free Press).

Meyer-Krahmer, F. (1997), 'Basic Research for Innovation: The Case of Science-Based Technologies', in P. Shearmur, B. Osmond, and P. Pockley (eds.), *Nurturing Creativity in Research* (Canberra: Institute of Advanced Studies).

—— (1999), *Globalization of R & D and Technology Markets* (Heidelberg: Springer-Verlag).

Michael, D. (1973), *On Learning to Plan—and Planning to Learn* (San Francisco: Jossey-Bass).

Miles, I. (1994), 'Innovation in Services', in M. Dodgson and R. Rothwell (eds.), *The Handbook of Industrial Innovation* (Aldershot: Edward Elgar).

Miles, R., and Snow, C. (1994), *Fit, Failure and the Hall of Fame* (New York: Free Press).

Mintzberg, H. (1987), 'Crafting Strategy', *Harvard Business Review*, (July–Aug.), 66–75.

—— (1994), *The Rise and Fall of Strategic Planning* (New York: Free Press).

Mitchell, G. (1986), 'New Approaches for the Strategic Management of Technology', in M. Horwitch (ed.), *Technology in the Modern Corporation: A Strategic Perspective* (New York: Pergamon).

Miyazaki, K. (1995), *Building Competencies in the Firm: Lessons for Japanese and European Optoelectronics* (Basingstoke: Macmillan).

Mody, A. (1990), *Learning through Alliances* (Washington: World Bank).

Morgan, G. (1986), *Images of Organization* (Beverly Hills, Calif.: Sage).

Morris, C., and Ferguson, C. (1993), 'How Architecture Wins Technology Wars', *Harvard Business Review* (Mar.–Apr.), 86–95.

Mowery, D. (1987), *Alliance Politics and Economics: Multinational Joint Ventures in US Manufacturing* (Cambridge, Mass.: Ballinger).

—— (1988), *International Collaborative Ventures in US Manufacturing* (Cambridge, Mass.: Ballinger).

Mytelka, L. (1991*a*), 'States, Strategic Alliances and International Oligopolies: The European ESPRIT Programme', in L. Mytelka (ed.), *Strategic Partnerships and the World Economy* (London: Pinter Publishers).

—— (1991*b*), 'Crisis, Technological Change and the Strategic Alliance', in L. Mytelka (ed.), *Strategic Partnerships and the World Economy* (London: Pinter Publishers).

Narin, F., and Beitzmann, A. (1995), 'Inventive Productivity', *Research Policy*, 24: 507–19.

Narin, F. and Noma, E. (1985), 'Is Technology Becoming Science?', *Scientometrics*, 7/3–6: 369–81.

Nelson, R. (1993) (ed.), *National Innovation Systems: A Comparative Analysis* (New York: Oxford University Press).

—— and Winter, S. (1982), *An Evolutionary Theory of Economic Change* (Cambridge, Mass.: Belknap Press).

Nevis, E., DiBella, A., and Gould, J. (1997), 'Understanding Organizations as Learning Systems', *Sloan Management Review* (Winter), 73–85.

Newman, W. (1992), 'Focused Joint Ventures in Transforming Economies, *Executive*, 6/1: 67–75.

Newton, D., and Pearson, A. (1994), 'Application of Option Pricing Theory to R & D', *R & D Management*, 24: 83–9.

—— Paxson, D., and Pearson, A. (1996), 'Real R & D Options', in A. Belcher, J. Hassard, and S. Procter (eds.), *R & D Decisions: Strategy, Policy and Innovations* (London: Routledge).

Nicolson, G. (1998), 'Innovation—A Survival Issue', paper presented at the Innovation and Collaboration Conference, Taipei, 21–2 Jan.

Nishiguchi, T. (1994), *Strategic Industrial Sourcing: The Japanese Advantage* (New York: Oxford University Press).

Nonaka, I., and Kenney, M. (1991), 'Towards a New Theory of Innovation Management: A Case Study Comparing Canon, Inc. and Apple Computer Inc.', *Journal of Engineering and Technology Management*, 8: 67–83.

—— and Takeuchi, H. (1995), *The Knowledge-Creating Company* (New York: Oxford University Press).

NSF (1996): National Science Foundation, *Science and Engineering Indicators 1996* (Washington: National Science Foundation).

—— (1998), *Science and Engineering Indicators 1998* (Washington: National Science Foundation).

Odagiri, H. (1985), 'Research Activity, Output Growth, and Productivity Increase in Japanese Manufacturing Industries', *Research Policy*, 14/3: 117–30.

—— (1992), *Growth through Competition: Competition through Growth* (Oxford: Oxford University Press).

OECD (1992): Organization for Economic Cooperation and Development, *Technology and the Economy* (Paris: OECD).

—— (1994), *Accessing and Expanding the Science and Technology Base* (Paris: DSTI/STP/TIP).

—— (1996), *Employment and Growth in the Knowledge-Based Economy* (Paris: OECD).

—— (1997), *National Innovation Systems* (Paris: OECD).

—— (1998), *The Economic and Social Impact of Electronic Commerce* (Paris: OECD).

—— (1999), *The Knowledge-Based Economy: A Set of Facts and Figures* (Paris: OECD).

Oliver, N., and Wilkinson, B. (1988), *The Japanization of British Industry* (Oxford: Blackwell).

Patel, P., and K. Pavitt (1991*a*), 'Europe's Technological Performance', in C. Freeman, M. Sharp, and W. Walker (eds.), *Technology and the Future of Europe* (London: Pinter Publishers).

—— —— (1991*b*), 'Large Firms in the Production of the World's Technology: An Important Case of Non-Globalisation', *Journal of International Business*, 22/1: 1–21.

—— —— (1994), 'National Innovation Systems: Why they are Important, and How they might be Measured and Compared', *Economics of Innovation and New Technology*, 3: 77–95.

—— —— (1998), 'The Wide (and Increasing) Spread of Technological Competencies in the World's Largest Firms', in A. Chandler, P. Hagstrom, and O. Solvell (eds.), *The Dynamic Firm* (Oxford: Oxford University Press).

Pavitt, K. (1986), 'Technology, Innovation, and Strategic Management', in J. McGee and H. Thomas (eds.), *Strategic Management Research* (Chichester: Wiley).

—— (1990), 'What We Know about the Strategic Management of Technology', *California Management Review*, 32/3: 17–26.

—— (1991), 'Key Characteristics of the Large Innovating Firm', *British Journal of Management*, 2: 41–50.

—— (1993), 'What Do Firms Learn From Basic Research', in D. Foray and C. Freeman (eds.), *Technology and the Wealth of Nations* (London: Macmillan).

—— (1994), 'Key Characteristics of Large Innovating Firms', in M. Dodgson and R. Rothwell (eds.), *The Handbook of Industrial Innovation* (Aldershot: Edward Elgar).

—— (1998), 'Technologies, Products and Organization in the Innovating Firm: What Adam Smith Tells Us and Joseph Schumpeter Doesn't', *Industrial and Corporate Change*, 7/3: 433–52.

—— and Patel, P. (1988), 'International Distributions and Determinants of Technological Activities', *Oxford Review of Economic Policy*, 4/4: 35–55.

—— Robson, M., and Townsend, J. (1989), 'Accumulation, Diversification and Organisation of Technological Activities in UK Companies, 1945–83', M. Dodgson (ed.), *Technology Strategy and the Firm* (Harlow: Longman).

Pearson, A. (1997), 'Innovation Management—Is There Still a Role for "Bootlegging"?', *International Journal of Innovation Management*, 1/2: 191–200.

Pisano, G. (1996), *The Development Factory: Unlocking the Potential of Process Innovation* (Boston, Mass.: Harvard Business School Press).

—— and Wheelwright, S. (1995), 'High-Tech R & D', *Harvard Business Review* (Sept.–Oct.), 93–105.

Polanyi, M. (1962), *Personal Knowledge: Towards a Post-Critical Philosophy* (New York: Harper & Row).

—— (1967), *The Tacit Dimension* (London: Routledge).

Porter, M. (1985), *Competitive Strategy* (New York: Free Press).

—— (1987), 'From Competitive Advantage to Corporate Strategy', *Harvard Business Review* (May–June), 43–59.

—— (1990), *The Competitive Advantage of Nations* (New York: Free Press).

Prahalad, C., and Hamel, G. (1990), 'The Core Competence of the Corporation', *Harvard Business Review* (May–June), 79–91.

—— —— (1994), *Competing for the Future* (Boston, Mass.: Harvard Business School Press).

Pucik, V. (1988*a*), 'Strategic Alliances with the Japanese: Implications for Human Resource Management', in F. Contractor and P. Lorange (eds.), *Cooperative Strategies in International Business* (Lexington, Mass.: Lexington Books).

—— (1988*b*), 'Strategic Alliances, Organisational Learning, and Competitive Advantage: The HRM Agenda', *Human Resource Management*, 27/1: 77–93.

Quinn, J. (1992), *Intelligent Enterprise* (New York: Free Press).

Reddy, N. (1987), 'Voluntary Products Standards: Linking Technical Criteria to Marketing Decisions', *IEEE Transactions on Engineering Management*, 34/4: 236–43.

—— Cort, S., and Lambert, D. (1989), 'Industrywide Technical Product Standards', *R & D Management*, 19/1: 13–25.

Reddy, P. (1997), 'New Trends in Globalization of Corporate R & D and Implications for Innovation Capability in Host Countries: A Survey from India', *World Development*, 25/11: 1821–37.

Reger, G. (1997), 'Benchmarking the Internationalization and Co-ordination of R & D of Western European and Japanese Multinational Corporations', *International Journal of Innovation Management*, 1/3: 299–331.

Roberts, E. (1991), *Entrepreneurs in High Technology: Lessons from MIT and Beyond* (Oxford: Oxford University Press).

—— (1994), *Benchmarking the Strategic Management of Technology*, i (Cambridge, Mass.: MIT Sloan).

—— (1995), *Benchmarking the Strategic Management of Technology*, ii (Cambridge, Mass.: MIT Sloan).

Roehl, T., and Truitt, J. (1987), 'Stormy Open Marriages are Better: Evidence from US, Japanese and French Cooperative Ventures in Commercial Aircraft', *Columbia Journal of World Business* (Summer), 87–95.

Romer, P. (1990), 'Endogenous Technical Change', *Journal of Political Economy*, 98/5: 71–102.

Rosenberg, N. (1982), *Inside the Black Box: Technology and Economics* (Cambridge: Cambridge University Press).

—— (1990), 'Why do Firms do Basic Research (With their own Money)?', *Research Policy*, 19/2: 165–75.

Rothwell, R. (1992), 'Successful Industrial Innovation: Critical Factors for the 1990s', *R & D Management*, 22/3: 221–39.

—— (1994), 'Industrial Innovation: Success, Strategy, Trends', in M. Dodgson and R. Rothwell (eds.), *The Handbook of Industrial Innovation* (Aldershot: Edward Elgar).

—— and M. Dodgson (1994), 'Innovation and Size of Firm', in M. Dodgson and R. Rothwell (eds.), *The Handbook of Industrial Innovation* (Aldershot: Edward Elgar).

—— and Gardiner, P. (1988), 'Re-Innovation and Robust Designs: Producer and User Benefits', *Journal of Marketing Management*, 3/3: 372–87.

—— and Zegveld, W. (1985), *Reindustrialization and Technology* (Harlow: Longman).

—— Freeman, C., Horley, A., Jervis, V., Robertson, Z., and Townsend, J. (1974), 'SAPPHO Updated—Project SAPPHO, Phase II', *Research Policy*, 3: 258–91.

Roussel, P., Saad, K., and Erickson, T. (1991), *Third Generation R & D* (Boston, Mass.: Harvard Business School Press).

Rubenstein, A. (1989), *Managing Technology in the Decentralized Firm* (New York, Wiley).

—— Geisler, E., and Grabowski, B. (1994), 'Managing Technology in the Service Industries' in E. Rhodes and D. Wield (eds.), *Implementing New Technologies* (Oxford: NCC Blackwell).

Sabbagh, K. (1996), *Twenty-First-Century Jet: The Making and Marketing of the Boeing 777* (New York: Scribner).

Sabel, C. (1993), 'Studied Trust: Building New Forms of Cooperation in a Volatile Economy', in D. Foray and C. Freeman (eds.), *Technology and the Wealth of Nations*, (London: Pinter).

Sako, M. (1992), *Prices, Quality and Trust: How Japanese and British Companies Manage Buyer–Supplier Relations* (Cambridge: Cambridge University Press).

—— and Helper, S. (1998), 'Determinants of Trust in Supplier Relations: Evidence from the Automotive Industry in Japan and the United States', *Journal of Economic Behaviour and Organization*, 34: 387–417.

Sampson, A. (1995), *Company Man* (London: HarperCollins).

Sanderson, S., and Uzumeri, M. (1995), 'Managing Product Families: The Case of the Sony Walkman', *Research Policy*, 24/5: 761–82.

Sato, Y. (1998), 'The Transfer of Japanese Management Technology to Indonesia', in H. Hill and T. K. Wie (eds.), *Indonesia's Technological Challenge* (Canberra: Australian National University).

Saxenian, A. (1991), 'The Origins and Dynamics of Production Networks in Silicon Valley', *Research Policy*, 20: 423–37.

—— (1994), *Regional Advantage* (Boston, Mass.: Harvard Business School Press).

Schein, E. (1985), *Organizational Culture and Leadership* (San Francisco: Jossey-Bass).

Schonberger, R. (1983), *Japanese Manufacturing Techniques* (New York: Free Press).

Senge, P. (1990), 'The Leader's New Work: Building Learning Organizations', *Sloan Management Review*, 32/1: 7–23.

Senker, J., Joly, P., and Reinhard, M. (1996), *Overseas Biotechnology Research by Europe's Chemicals/Pharmaceuticals Multinationals: Rationale and Implications* (Brighton: Science Policy Research Unity).

Sheehan, P., and Tikhomirova, G. (1998), 'The Rise of the Global Knowledge Economy', in P. Sheehan and G. Tegart (eds.), *Working for the Future: Technology and Employment in the Global Knowledge Economy* (Melbourne: Victoria University Press).

—— Pappas, N., Tikhomirova, G., and Sinclair, P. (1995), *Australia and the Knowledge Economy* (Melbourne: Centre for Strategic Economic Studies).

Sigurdson, J. (1998), personal communication.

—— (1999), personal communication.

Skinner, W. (1985), *Manufacturing, the Formidable Competitive Weapon* (New York: Wiley).

Smith, G., Lederman, F., and Jonash, R. (1999), 'Alcoa's Technology Change Process', mimeo.

Soete, L. (1991), *Policy Synthesis* (OECD Technology Economy Programme, MERIT; University of Limburg, Maastricht).

Starbuck, W. (1992), 'Learning by Knowledge-Intensive Firms', *Journal of Management Studies*, 29: 713–40.

Steinmueller, E. (1994), 'Basic Research and Innovation', in M. Dodgson and R. Rothwell (eds.), *The Handbook of Industrial Innovation* (Aldershot: Edward Elgar).

Stevens, C. (1996), 'The Knowledge-Driven Economy', *OECD Observer*, 200: 6–10.

Subramanian, S. (1990), 'Managing Technology—the Japanese Approach', *Journal of Engineering and Technology Management*, 6: 221–36.

Sveiby, K. (1997), *The New Organizational Wealth* (San Francisco: Berrett-Koehler).

Sykes, R. (1998), 'How the Sarich Vision Stalled', *Australian Financial Review Magazine*, 22–31.

Szakonyi, R. (1990), 'Coordinating R & D and Business Planning', *Technology Analysis and Strategic Management*, 2/4: 391–409.

Teece, D. (1986), 'Profiting from Technological Innovation: Implications for Integration, Collaboration, Licensing and Public Policy', *Research Policy*, 15: 285–305.

—— (1987) (ed.), *The Competitive Challenge* (Cambridge, Mass.: Ballinger).

—— (1998), 'Licensing and the Market for Know-how', *R & D Enterprise—Asia Pacific*, 1/2–3: 3–5.

—— and Pisano, G. (1994), 'The Dynamic Capabilities of Firms: An Introduction', *Industrial and Corporate Change*, 3/3: 537–56.

—— Pisano, G., and Shuen, A. (1990), 'Firm Capabilities, Resources, and the Concept of Strategy: Four Paradigms of Strategic Management, (Working Paper, Berkeley University, San Francisco).

Teubal, M. (1996), 'R & D and Technology Policy in NICs as Learning Processes', *World Development*, 24/3: 449–60.

Thamhain, H., and Wilemon, D. (1987), 'Building High Performance Engineering Project Teams', *IEEE Transactions on Engineering Management*, 34/3: 130–7.

Thomke, S. (1997), 'The Role of Flexibility in the Development of New Products: An Empirical Study', *Research Policy*, 26/1: 105/20.

Tidd, J. (1991), *Flexible Manufacturing Technologies and International Competitiveness* (London: Pinter).

Turban, E., McLean, E., and Wetherbe, J. (1996), *Information Technology for Management* (New York: Wiley).

Tushman, M., and Anderson, P. (1986), 'Technological Discontinuities and Organizational Environment', Administrative Science Quarterly, 31: 439–56.

—— and O'Reilly, C. (1997), *Winning through Innovation* (Boston, Mass.: Harvard Business School Press).

—— Virany, B. and Romanelli, E. (1986), 'Executive Succession, Strategic Reorientation, and Organizational Evolution', in M. Horwitch (ed.), *Technology in the Modern Corporation: A Strategic Perspective* (New York: Pergamon).

UK Department of Trade and Industry (1998), *Our Competitive Future Building the Knowledge Economy* (London: HMSO).

Upton, D. (1995), 'What Makes Factories Really Flexible?', *Harvard Business Review* (July–Aug.), 74–84.

US Department of Commerce (1994), *Technology, Economic Growth and Employment* (Washington: Office of the Chief Economist).

Utterback, J. (1994), *Mastering the Dynamics of Industrial Innovation* (Boston, Mass.: Harvard Business School Press).

Vaitsos, C. (1974), *Intercountry Income Distribution and Transnational Enterprise* (Oxford: Oxford University Press).

Vernon, R. (1966), 'International Investment and International Trade in the Product Cycle', *Quarterly Journal of Economics*, 80/2: 190–207.

—— (1979), 'The Product–Cycle Hypothesis in a New International Environment', *Oxford Bulletin of Economics and Statistics*, 41/4: 255–67.

Von Glinow, M., and Mohrman, A. (1990), *Managing Complexity in High Technology Organizations* (New York: Oxford University Press).

von Hippel, E. (1988), *The Sources of Innovation* (Oxford: Oxford University Press).

Voss, C. (1994), 'Implementation of Manufacturing Innovations', in M. Dodgson and R. Rothwell (eds.), *The Handbook of Industrial Innovation* (Aldershot: Edward Elgar).

Wakasugi, R. (1992), 'Why Are Japanese Firms So Innovative in Engineering Technology', *Research Policy*, 21: 1–12.

Wheelwright, S. (1988), 'Product Development and Manufacturing Start-Up', in M. Tushman and W. Moore (eds.), *Readings in the Management of Innovation* (New York: Harper Business).

Whiston, T. (1991), *Managerial and Organisational Integration* (Berlin: Springer-Verlag).

—— (1994), 'The Global Innovatory Challenge across the Twenty-First Century', in M. Dodgson and R. Rothwell (eds.), *The Handbook of Industrial Innovation* (Aldershot: Edward Elgar).

Whittaker, H. (1997), *Small Firms in the Japanese Economy* (Cambridge: Cambridge University Press).

Williamson, O. (1975), *Markets and Hierarchies: Analysis and Anti-trust Implications* (New York: Free Press).

—— (1985), *The Economic Institutions of Capitalism* (New York: Free Press).
Williamson, P., and Hu, Q. (1994), *Managing the Global Frontier* (London: Pitman).
Willoughby, K., and Wong, B. (1993), 'Orbital Engine Company Pty Ltd', in G. Lewis, A. Morkel, and G. Hubbard (eds.), *Australian Strategic Management: Concepts, Context and Cases* (Sydney: Prentice Hall).
Womack, J., and Jones, D. (1996), *Lean Thinking* (New York: Simon & Schuster).
—— —— Roos, D. (1990), *The Machine that Changed the World* (New York: Rawton).
Wong, P.–K., and Mathews, J. (1998), 'Competing in the Global Flat Panel Display Industry', *Industry and Innovation*, 5/1: 1–10.
World Bank (1998), *World Development Report 1998/99* (Washington: World Bank).
Wyatt, S. (1984), *The Role of Small Firms in Innovatory Activity* (Brighton: Science Policy Research Unit).
Zangwill, W. (1993), *Lighting Strategies for Innovation* (New York: Lexington).

INDEX